The Minerals, Metals & Materials Series

T. S. Srivatsan · Manoj Gupta
Editors

Nanocomposites VI: Nanoscience and Nanotechnology in Advanced Composites

TMS

Springer

Editors
T. S. Srivatsan
Department of Mechanical Engineering
Auburn Science and Engineering Center
The University of Akron
Akron, OH, USA

Manoj Gupta
Department of Mechanical Engineering
National University of Singapore
Singapore, Singapore

ISSN 2367-1181 ISSN 2367-1696 (electronic)
The Minerals, Metals & Materials Series
ISBN 978-3-030-35789-4 ISBN 978-3-030-35790-0 (eBook)
https://doi.org/10.1007/978-3-030-35790-0

Cover illustration:
Top Transmission electron micrograph showing two different morphologies [rod-shaped and blocky precipitates] of Mg-Zn-rich phase and the presence of Y_2O_3 nanoparticles in an extruded Mg-1.8Y/1.53ZnO nanocomposite.
Bottom Influence of reinforcement content (volume pct.) on tensile yield strength of magnesium alloy nanocomposites. [From research of Dr. Manoj Gupta group]

This Springer imprint is published by the registered company Springer Nature Switzerland AG
The registered company address is: Gewerbestrasse 11, 6330 Cham, Switzerland

Preface

The development, emergence, and increased focus in the selection and use of nanocomposites has in more recent years engendered considerable scientific and technological interest as potentially viable and affordable replacements to not only the heavier and lighter monolithic metals but also their conventional composite counterparts, i.e., the family of metal-matrix composites, for the purpose of selection and use in a spectrum of both performance-critical and non-performance-critical applications. A high demand and interest in these composites have been particularly seen in the industries specific to air transportation, ground transportation, high-performance defense-related products, and even commercial products to include electronics and sporting goods. In an attempt to explore the prevailing state of knowledge and aspects specific to research, advancement, and growth by way of selection and use in appropriate applications and to concurrently attempt to bring forth aspects specific to innovation, advances and applications in the domains spanning both nanoscience and nanotechnology for the purpose of engineering the development and use of these "novel" materials, this international symposium, the sixth in a series, was held with the primary objective of bringing to light aspects specific to recent and novel innovations, advances, and applications of the emerging family of nanocomposites. Appropriately, this symposium was given the subtitle "Nanoscience and Nanotechnology in Advanced Composites" and was held at the TMS 2020 149th Annual Meeting and Exhibition in San Diego, California, USA (February 23–27, 2020). Overall, this symposium was well represented with abstracts from both the United States and several nations around the world and from engineers and scientists associated with the sectors of academia, research institutions, and industry. This volume contains papers presented at the four-session symposium that was approved by both the Composite Materials Committee and Structural Materials Division of TMS. Over 30 abstracts were approved for presentation in the sessions:

Session 1: Processing and Fabrication of Nanocomposites
Session 2: Microstructure and Properties of Nanocomposites
Session 3: Processing–Properties–Performance of Nanocomposites
Session 4: Polymer and Other Nanocomposites

In the four oral sessions, recent advances specific to engineering the development, characterization, rationalization, and properties of the family of nanocomposites were presented. It was clear that every presenter did make an effort to present and discuss new and/or emerging knowledge very much in tune with his ideas. We, the symposium organizers, extend our warmest thanks and appreciation to both the presenters and the participants. We also extend our sincere thanks and much deserving appreciation to the elected and governing representatives of the Composite Materials Committee and Structural Materials Division of TMS for their valued understanding and timely approval of all related and relevant paperwork specific to this symposium.

Our gracious and generous gratitude coupled with lasting appreciation and recurring thanks are also extended to the following individuals positioned strategically at the head office of TMS (Pittsburgh, Pennsylvania, USA):

(a) **Ms. Trudi Dunlap** (Department Head, Events, Education, and Exhibitions) for her early interest and involvement by way of observable and much needed attention and assistance. Ms. Dunlap ensured a smooth transition to Ms. Patricia Warren for the purpose of timely execution of all related and relevant intricacies, numerous in number, specific to orchestration, layout, and conduct of the symposium from the moment following its approval all the way until the end signified by compilation and publication of this volume.
(b) An unbounded dose of recurring thanks and highly deserving appreciation, of resonating order, is reserved and extended to **Ms. Patricia Warren** (Programming and Proceedings Specialist) for her infinite patience, valued understanding, and the much desired and needed attention to all of the specifics and intricacies, far too many to list, from point of conception to completion of this technical event by way of compilation and presentation of this impressive bound volume.
(c) The timely publication of this volume was made possible by the cooperation of the contributing authors and the publishing staff headed by the dedicated and diligent **Mr. Matt Baker** (Department Head, Content).

We the editors have every confidence that this volume will provide all the readers, to include both the experienced and the new entrant, both new perspectives and directions in their research efforts in the domain specific to nanocomposites. It is anticipated that the sustained growth and resultant demand for nanocomposites will only show a continuing upward trend. Thus, by making possible this symposium we have made every effort to not only aid but also assist material developers in making every possible effort in engineering the safe development of nanocomposites to ensure their eventual selection and use in few to many potential applications.

Dr. T. S. Srivatsan
Professor

Dr. Manoj Gupta
Associate Professor

Contents

About the Editors

T. S. Srivatsan is Professor (Emeritus) in the Department of Mechanical Engineering at The University of Akron. He received his graduate degrees—Master of Science in Aerospace Engineering (M.S. 1981) and Doctor of Philosophy in Mechanical Engineering (Ph.D. 1984)—from Georgia Institute of Technology. Dr. Srivatsan joined the faculty in the Department of Mechanical Engineering at The University of Akron in August 1987. Over the years he served both at The University of Akron and Georgia Institute of Technology, and he has taught courses in the areas spanning Materials Science and Engineering, Manufacturing Processes, Fracture Mechanics, Fatigue of Engineering Materials and Structures, Mechanical Behavior of Materials, Mechanical Measurements, and Mechanical Design. His research areas have spanned studies investigating and understanding the fatigue and fracture behavior of advanced materials to include monolithic(s), intermetallic, nanomaterials and metal-matrix composites; processing techniques for advanced materials and nanostructure materials; inter-relationship between processing and mechanical behavior; electron microscopy; failure analysis; and mechanical design. He has authored/edited/co-edited sixty books and five research monographs in areas cross-pollinating mechanical design; processing and fabrication of advanced materials; deformation, fatigue, and fracture of ordered intermetallic materials; machining of composites; metal-matrix composites; failure analysis; and the technology of rapid solidification processing of materials. He serves as co-editor of the *International Journal on Materials and Manufacturing Processes* and on the editorial advisory board of five journals in the domain of materials science and engineering and manufacturing processes. His research has enabled him to deliver more than 230 technical presentations in national and international meetings and symposia. He has authored and co-authored over 700 journal articles, book chapters, proceedings papers, book reviews, and technical reports. His RG score is 45 with an h-index of 51 and Google scholar citations of 8,500, ranking him to be among the top 2 percent of researchers in the world. In recognition of his efforts, contributions and its impact on furthering science, technology, and education he has been elected:

(a) Fellow of ASM International.
(b) Fellow of American Society of Mechanical Engineers.
(c) Fellow of American Association for the Advancement of Science.

He has also been recognized as: (i) Outstanding Young Alumnus of Georgia Institute of Technology; (ii) Outstanding Research Faculty, of the College of Engineering [The University of Akron]; and (iii) Outstanding Research Faculty at The University of Akron. He offers his knowledge and services to the National Research Laboratories, and industries related to aerospace, automotive, power generation, leisure-related products, and applied medical sciences.

Manoj Gupta is the former Head of Materials Division of the Mechanical Engineering Department and Director Designate of Materials Science and Engineering Initiative at NUS, Singapore. He completed his Ph.D. at the University of California, Irvine, USA (1992), and did postdoctoral research at the University of Alberta, Canada (1992). In August 2017 he was highlighted as a Top 1% Scientist of the World Position by The Universal Scientific Education and Research Network and among 2.5% among scientists as per ResearchGate. To his credit are: (i) disintegrated melt deposition technique and (ii) hybrid microwave sintering technique, an energy-efficient solid-state processing method to synthesize alloys/micro/nanocomposites. He has published over 535 peer-reviewed journal papers and owns two U.S. patents and one trade secret. His current h-index is 61, RG index is > 47, and citations are greater than 14,000. He has also co-authored six books, published by John Wiley, Springer, and MRF-USA. He has served as editor-in-chief/editor of 12 international peer-reviewed journals. In 2018 he was announced World Academy Championship Winner in the area of biomedical sciences by the International Agency for Standards and Ratings. A multiple award winner, he actively visits Japan, France, Saudi Arabia, Qatar, China, India, and the United States as a visiting researcher and professor.

Part I
Processing and Fabrication of Nanocomposites

Processing, Properties and Potential Applications of Magnesium Alloy-Based Nanocomposites: A Review

Sravya Tekumalla and Manoj Gupta

Abstract Strong, ductile, lightweight, biocompatible and non-toxic materials are the need of the hour for metal-based industries such as aerospace, automotive, electronics and biomedical sectors. Magnesium-based materials, due to their lightweight, excellent dimensional stability and mechanical integrity, have a tremendous potential to replace the existing commercial Al, Ti alloys and steels currently being used. Due to these attractive qualities of magnesium, there has been a spurt in the quest of a variety of magnesium materials targeting different functionalities. One subset of magnesium-based materials is magnesium alloy-based nanocomposites that exhibit advantages of both magnesium alloys and magnesium nanocomposites. There has been advancement in this field through careful selection of alloying elements and reinforcement and optimization to obtain the best combination of properties. Accordingly, this paper will focus on the recent developments of magnesium alloy-based nanocomposites capable of replacing conventional materials in multiple engineering and biomedical applications.

Keywords Magnesium · Alloy nanocomposites · Microstructure · Properties

Introduction

There has been an ever-increasing trend in catastrophic changes in the weather patterns that are being recorded ever since the beginning of 1990s [1]. These are primarily attributed to a progressive increase in greenhouse gas emissions along with several other factors such as increase in the toxicity in land and water bodies. The extent of these damages is increasing at an alarming rate never seen before in the

S. Tekumalla (✉)
School of Mechanical and Aerospace Engineering, Nanyang Technological University, Singapore 639798, Singapore
e-mail: tvrlsravya@u.nus.edu

S. Tekumalla · M. Gupta
Department of Mechanical Engineering, National University of Singapore, Singapore 117575, Singapore

T. S. Srivatsan and M. Gupta (eds.), *Nanocomposites VI: Nanoscience and Nanotechnology in Advanced Composites*, The Minerals, Metals & Materials Series, https://doi.org/10.1007/978-3-030-35790-0_1

history of mankind. In this view, several guidelines are laid out such as keeping the temperature rise to within 2 °C with respect to preindustrial times and to limit the emissions by 2025 to 2 billion tons [2]. While this is an example of the catastrophic problems happening at a global scale, there are also several challenges that a man faces in everyday life such as emergence of electromagnetic smog due to spike in the use of electronic gadgets which poses a higher health risk. Furthermore, with the increase in the ageing population at an unprecedented rate along with a need for materials in biomedical applications that help the patients recover faster without having any effects of toxicity in the human body, there is a need for a paradigm shift in the usage of materials that can help reach all the above said goals. In this view, magnesium serves as a phenomenal material with multiple roles to play. It was ignored to a large extent so far and that ignorance is, in part, responsible for the worsening climate and contamination of land mass and water bodies all over the globe. Magnesium, one of the most diverse metals, has its widespread presence in planet earth and the universe besides being the fourth most abundant cation in human bodies. The key attributes of magnesium are its lightweight (~35% lighter than aluminium, ~75% lighter than steel), its abundance, non-toxicity (a nutritional element for both plants and humans [3]) and most importantly, its recyclability. Over the last 15 years, the price of magnesium has been showing a downward trend making it an increasingly cost-effective option in several applications. Magnesium based nanocomposites have been in the literature for more than a decade as promising materials for aerospace and automotive applications [4]. However, due to their relatively poor ignition and corrosion resistance, and scalability, they have not been explored much commercially [5]. In the recent past, there have been several articles reporting the improved mechanical and ignition response of magnesium with the combined addition of alloying elements and nanoparticles [2, 6–8]. Therefore, it is extremely meaningful to understand the behavior of magnesium alloy-based nanocomposites whose properties and performance depend on both the alloying elements added as well as the reinforced nanoparticles. The multifaceted potential of these magnesium alloy nanocomposites will be unravelled in detail in this paper, and special focus will be given to the engineering and biomedical applications of these materials.

Processing

Some of the commonly employed processing techniques to synthesize magnesium alloy nanocomposites are reviewed and discussed below.

Disintegrated Melt Deposition

Disintegrated melt deposition (DMD) is a unique processing method which inherits the combined advantages of stir casting and spray processing techniques [4]. This method involves the vortex mixing of ceramic reinforcements into the matrix material and deposition of the resulting molten slurry onto a metallic substrate after disintegration by jets of inert gases. Unlike conventional spray processes, this method employs higher superheat temperatures and lower impinging velocity which helps in achieving a bulk composite material with no-over sprayed powders. Hence, this process offers both the features of (i) fine grain structure and low segregation of reinforcements of spray process and (ii) simplicity and cost-effectiveness of conventional stir cast foundry process. Magnesium-based materials are synthesized by melting and casting commercially pure Mg turnings and other alloying/reinforcement additions. This method involves adding the raw material in alternate layers to form a sandwich pattern in a graphite crucible and heating it in an electrical resistance furnace to 750 °C in a protective inert argon gas atmosphere. This method employs vortex stirring of melt at 450 rpm for 5 min. The stirrer used is a mild steel impeller with twin blade (pitch 45°) coated with Zirtex 25 to avoid contamination of molten metal with iron. The melt, released through an orifice of 10 mm diameter, located at the crucible's base was disintegrated by two argon gas jets that were oriented normal to the melt stream to obtain near equiaxed grain structure. The disintegrated melt was then deposited on the substrate forming an ingot of magnesium-based alloy nanocomposite (Fig. 1).

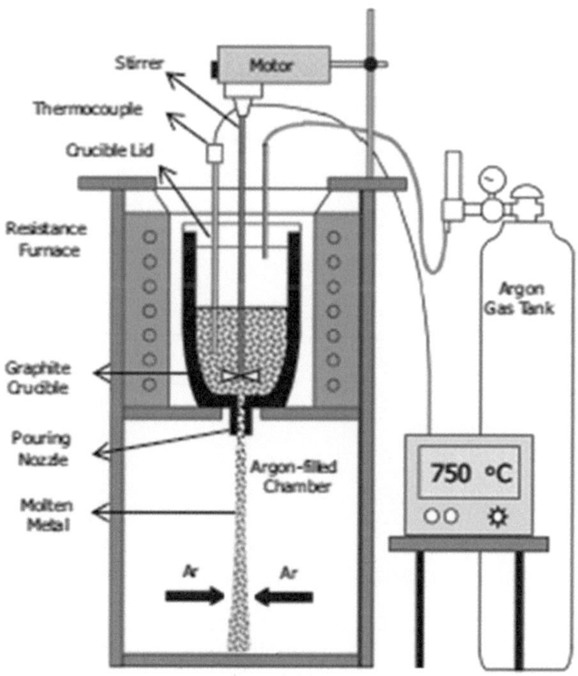

Fig. 1 Disintegrated melt deposition technique [4]

Ultrasound-Assisted Casting

Magnesium alloy nanocomposites with various matrix compositions, such as AZ31, Z6, ZC63, ZC71, and AZ91, have been produced using this method. Ultrasonic cavitation method [9] (Fig. 2a) has been used for improved reinforcement dispersion effects. It employs the transient (in the order of nanoseconds) micro-hot spots that can reach temperatures of about 5000 °C, pressures above 1000 atms, and heating and cooling rates above 1010 °C/s. Such strong impact coupling with local high temperatures will potentially break the nanoparticle clusters and clean the particle surface. SiC nanoparticles reinforced pure magnesium, Mg−4Zn and Mg-(2,4)Al−1Si by ultrasonic cavitation based solidification processing were developed and the experimental findings were promising. Recently, a simple addition to the ultrasonic processing was demonstrated by Chen et al. [10] to synthesize uniformly dispersed nanocomposites containing up to 14 vol.% of nanoparticulates. SiC nanoparticles were introduced in the magnesium-zinc melt and dispersed by ultrasonic processing. After solidification, an ingot was obtained which was evaporated such that the magnesium and zinc melt leaving a high concentrate of SiC nanoparticles. Then the samples were cooled down to room temperature in the furnace. It is reported that SiC nanoparticles self-stabilize in magnesium melt. This is due to the (i) synergy of reduced van der Waals forces between the nanoparticles in molten magnesium, (ii) a high thermal energy of the nanoparticles, and (iii) a high energy barrier preventing nanoparticle from sintering owing to a reasonable wettability between nanoparticles and molten magnesium. The schematic of ultrasonic processing and the evaporation of the melt are shown in Fig. 2b, c.

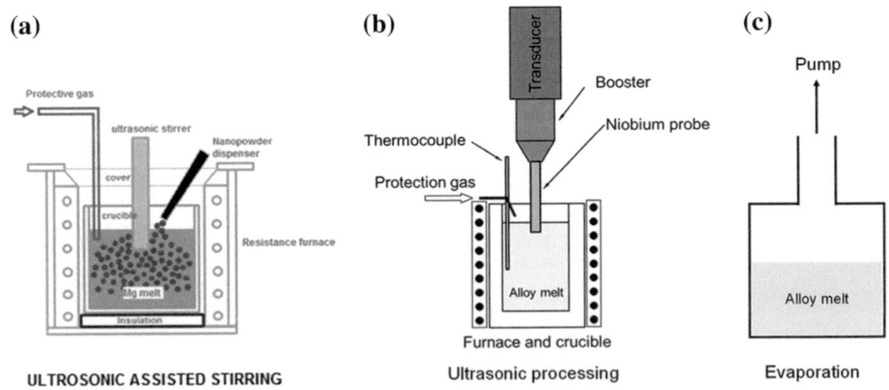

Fig. 2 Schematic of **a** ultrasonic-assisted stir casting setup [26]; **b** ultrasonic processing for nanoparticle feeding and dispersion, and **c** vacuum evaporation for concentrating nanoparticles in magnesium as demonstrated by Chen et al. [10]

In Situ Synthesis

The in situ synthesis method refers to the process in which the nanoparticles are synthesized during the casting process as a result of the reaction between the raw materials. It is a very common synthesis technique in aluminium-based materials; however, there has been very little reported work on magnesium composites processed through this technique. This can be achieved through (i) gas–liquid, (ii) liquid–liquid, and (iii) solid–liquid reactions based on phase diagram principles. Magnesium reinforced by in situ particulates was successfully synthesized using the reaction between pure magnesium and salts, viz., potassium boron tetrafluoride (KBF_4), and potassium titanium hexafluoride (K_2TiF_6) compounds resulting in titanium borides reinforcement particles [11]. Recently, an article [5] suggested in situ synthesizing and processing of magnesium alloy nanocomposites through the in situ evolution of the nanoparticles in the matrix. In this approach, the thermodynamic reactions between the matrix and reinforcements are to be considered and the kinetics play a crucial role in the in situ evolution. For example, in the Mg−1.8Y/1.53ZnO nanocomposite, Mg−1.8Y alloy was used as the matrix with ZnO nanoparticles added as reinforcements. During the processing, owing to the thermodynamic reactions, the ZnO nanoparticles dissolution induced the in situ formation of nano length scale rod-shaped Mg–Zn β_1' phase in the matrix. Oxygen released in the reaction forms Y_2O_3 nanoparticles in Mg matrix as shown in Fig. 3. Moreover, the β_1' phase and Y_2O_3 nanoparticles identified under TEM were shown to range between several tens and hundreds of nanometers. It is analyzed that the kinetics during melt processing at 750 °C led to the in situ formation of β_1' phase and Y_2O_3. This in situ evolution contributed to the grain size reduction and strengthening of the nanocomposite by improving its yield strength by

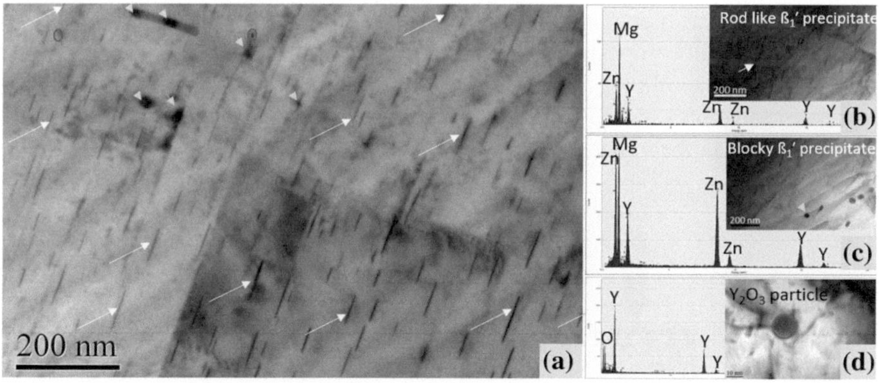

Fig. 3 **a** TEM image indicating the two different morphologies (rod-shaped and blocky precipitates) of Mg–Zn-rich phase and presence of Y_2O3 nanoparticles found in an extruded Mg−1.8Y/1.53ZnO nanocomposite; TEM image and EDS spectra of **b** nano $ß_1'$ phase rods, **c** blocky $ß_1'$ precipitates, and **d** Y_2O_3 nanoparticle

124% compared to the monolithic alloy. This mechanism has since then been proposed for in situ fabrication of magnesium matrix nanocomposites by optimizing the combination of alloying and nano-reinforcement additions that satisfy the thermodynamic criterion to obtain maximized benefit in terms of the properties. Although there has been some exploratory work on in situ synthesis of reinforcements through the use of different reinforcements, there has been very limited amount of work on this subject [12, 13] and several other magnesium alloy nanocomposites, processed through this route, can be explored and studied.

Blend-Press-Sinter Using Hybrid Microwave Sintering

Generally, the solid phase processes yield better strength properties because of minimal brittle interface reaction products. The major solid phase processing methods available includes the blend-press-sinter method. Blend-press-sinter technique is one of the conventional and well-established solid-state processing methods for the production of metal matrix composites. The schematic of Blend-Press-Sinter powder metallurgy method is shown in Figure. The first step involves blending of metal powder together with the reinforcement particles. After blending, consolidation of blended powder mixture is done to obtain a green compact. It is then forged or extruded. The advantage of PM method includes the capability of using different types of reinforcements and possibility of higher volume fraction of reinforcements. The major disadvantage is that this process cannot be used to produce complex shapes. The other limitations include the difficulty in removal of the binder which is normally used to hold the metal powder and reinforcement particles together and the inherent danger in handling the highly reactive powder used in the process (Fig. 4).

After the above-mentioned primary processing techniques, typically the magnesium alloy nanocomposites have been processed using secondary processing such as extrusion, rolling, forging, etc. This is to reduce the porosity and obtain finer microstructures and to increase the homogeneity of the second phases and/or reinforcements in the materials.

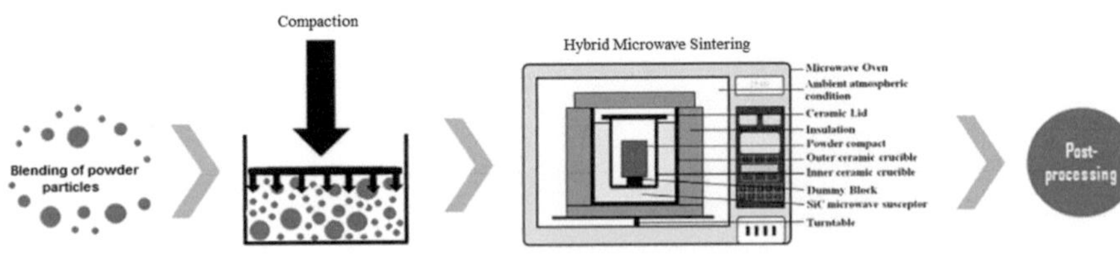

Fig. 4 Blend-press-sinter (hybrid microwave sintering) process flow chart

Properties

Mechanical Properties

Tensile Properties

In recent years, there has been a renewed interest on nanoparticle reinforced magnesium metal matrix composites. A study of available open literature indicates that the ceramic particles such as SiC, B_4C, TiC, Al_2O_3, Y_2O_3, ZrO_2, TiO_2, Bi_2O_3, Si_3N_4, TiN, BN, AlN, TiB_2 are used as nano-reinforcement for Mg and Mg alloys [4]. Owing to the inherent high temperature stability, high strength and high modulus of these ceramic phases, the dispersion of only a small volume percent in Mg offers distinct advantages through a combination of different strengthening mechanisms such as grain refinement strengthening, Orowan strengthening, geometrically necessary dislocations (GND) assisted strengthening effects, etc., without adversely affecting the ductility. Figure 5 gives the effect of reinforcements on various magnesium alloy matrices. It is seen that excellent strengths can be achieved through the usage of reinforcements to alloys due to their tremendous load-bearing potentials. Thus, excellent tensile behavior has been observed for alloy nanocomposites without compromise in their ductility.

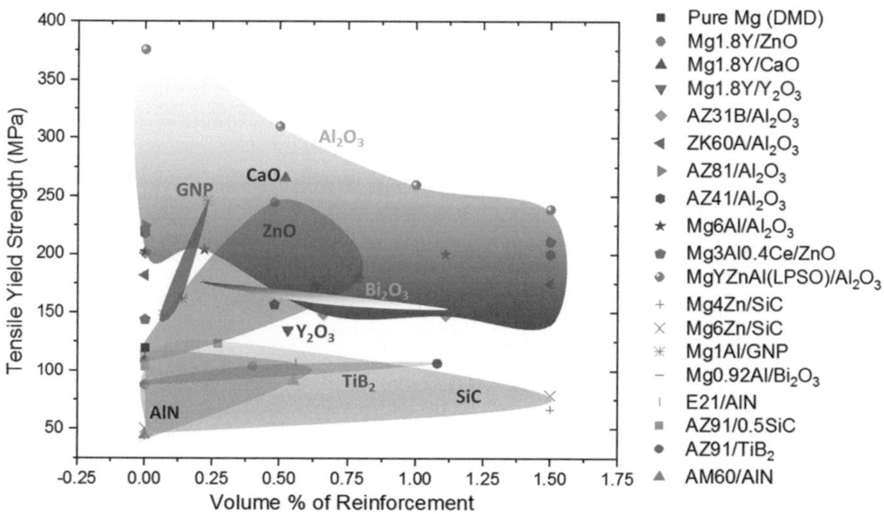

Fig. 5 Tensile yield strengths of magnesium alloy nanocomposites [4]

Compressive Properties

Similar to the tensile behavior, magnesium alloy nanocomposites exhibited tremendous properties under compression. The average compressive yield strengths are given in Table 1. One significant finding was by Chen et al. [14] where addition of ZnO nanoparticles enhanced the compressive strengths of alloy by ~100 MPa. The Mg−4Zn−3Gd−1Ca–ZnO [14] alloy nanocomposite samples display high strength and good ductility: 0.2% compressive yield stress of 355 MPa, ultimate compressive stress of 703 MPa, and compressive failure strain of 10.6%. The significant enhancement of compressive yield stress is mainly attributed to the grain refinement by adding nanoparticles. The strength levels exceed that of commercial magnesium alloys (i.e. WE43, WE54, ZK60, and ME21) and mild steels (i.e. S275 and S355), making

Table 1 Compressive properties of magnesium alloy matrix nanocomposites [4]

Material	Average compressive yield strength (MPa)
Pure Mg	66
AZ31B	133
AZ31B/0.66Al_2O_3	172
AZ31B/1.11Al_2O_3	174
AZ31B/1.50Al_2O_3	176
ZK60A	93
ZK60A/1.5 vol.% Al_2O_3	89
AZ81	157
AZ81/1.5 vol.% Al_2O_3	146
Mg–3Al–2.5La [8]	141
Mg–3Al–2.5La–1.9Y_2O_3	156
Mg−4Zn−3Gd−1Ca–ZnO [14]	**355**
Mg/0.92Al–0.22Bi_2O_3	140
Mg/0.92Al–0.66Bi_2O_3	143
Mg/0.92Al–1.11Bi_2O_3	115
Mg–3Al–0.1GNP	118
Mg–3Al–0.3GNP	134
Mg–3Al–0.5GNP	116
Mg–1.8Y	109
Mg–1.8Y–1.53ZnO [5]	160
Mg–1.8Y/1CaO [2]	111
Mg–1.8Y–1.5Y_2O_3 [15]	131
E21	92.4
E21-1AlN	88

Table 2 Coefficient of thermal expansion of magnesium, commonly used reinforcements [17] and a few magnesium alloy nanocomposites

Material	Coefficient of thermal expansion (μm/m.K)
Mg	25–27
ZnO	2.9
Y_2O_3	7.2
SiC	3.7
Al_2O_3	7.2
ZrO_2	10.5
Mg–3Al–0.4Ce	25.52 ± 0.8
Mg–3Al–0.4Ce–1.5ZnO	**22.03 ± 1.0**

Mg−4Zn−3Gd−1Ca/2ZnO a very promising material for multiple engineering and biomedical applications.

High Temperature-Related Properties

Coefficient of Thermal Expansion

A key feature in application of metal matrix nanocomposites in the automotive engineering is dimensional stability over a predicted lifetime at ambient and elevated temperatures. It is established that the dimensional stability is higher if the coefficient of thermal expansion of the material is lower [16]. Thermal expansion refers to the tendency to change shape in response to a change in temperature, occurring through heat transfer. The lesser the CTE value, the lesser the tendency for the material to change shape when the temperature is increased. Thus, a material with lower CTE values is a thermally stable material. Typically, magnesium and its alloys exhibit a coefficient of thermal expansion of 25–27 μm/m.K. With the addition of nano-reinforcements, particularly ceramics, the coefficient of thermal expansion alters significantly. Table 2 gives a list of values of commonly used ceramic reinforcements for magnesium. Further, Table 1 also gives the coefficient of thermal expansion values of magnesium alloy-based nanocomposites and it can be seen that just 1.5% of ZnO significantly lowered the CTE of the nanocomposite thereby rendering it more thermally stable.

Ignition Properties

With the synergistic effect of alloying and reinforcement additions, the ignition behavior of the magnesium alloy nanaocomposites completely changes. The nano-reinforcement (typically, ceramics) provides insulating effect that affects the thermal conductivity of the nanocomposite and reduces heat absorption of the reinforcement that affects the specific heat of the nanocomposite. This aids in enhancing the ignition temperature, however, it is to be noted that this enhancement in ignition temperature is only to an extent. Another crucial factor for improving the resistance to ignition is the nature of the oxide layer formed on the metal surface [18]. This is governed by Pilling–Bedworth ratio [19] which is ratio of the volume of the elementary cell of a metal oxide to the volume of the elementary cell of the corresponding metal (from which the oxide is created) and is given by the following equation:

$$\mathrm{R_{PB}} = \frac{V_{MO}}{V_M} \tag{1}$$

where R_{PB} is the Pilling–Bedworth ratio, V is the volume, MO corresponds to metal oxide and M corresponds to metal.

When $1 < R_{PB} < 2$, the oxide coating is passivating and provides a protecting effect against further surface oxidation. Metals like aluminium, titanium, chromium-containing steels fall in this category. Magnesium has $R_{PB} < 1$, which indicates that the oxide coating layer is too thin, likely broken and provides limited protective effect. When magnesium is exposed to air at elevated temperatures, due to its unstable oxides, the metal ignites rapidly even before reaching its melting temperature. Hence, modification of this oxide layer is equally essential to get an improved ignition behavior of magnesium [20]. The easiest method of modification of oxide layers is through alloying. Alloying with metals with $1 < R_{PB} < 2$ such as Y [21], satisfy the criteria of enhancing the ignition temperature due to the formation of a protective and stable oxide layer adhering to the surface of the alloy. Thus, with the synergistic effects of alloying and reinforcement with ceramic particles, a high ignition temperature can be achieved in magnesium alloy nanocomposites. Figure 6 shows the oxidation mechanism of a Mg−1.8Y/1CaO nanocomposite. The formation of such protective oxides along with the presence of thermally insulating reinforcement led to the ignition temperature beyond 1000 °C in the nanocomposite.

Performance and Potential Applications

Automobile Industry

In automotive industry, reducing the weight by 10% leads to an improvement in fuel economy by 7% in a vehicle [22]. Besides being light in weight, magnesium alloys also demonstrate superior specific strength and excellent castability with other advantages given in Fig. 1, which make them promising materials for the automotive industry. Using magnesium alloy nanocomposites as alternative to the currently used materials, results in a weight reduction by 22–70%. Among several magnesium alloys, Mg–Al-based alloy series, i.e. the AZ and AM alloys, demonstrate good

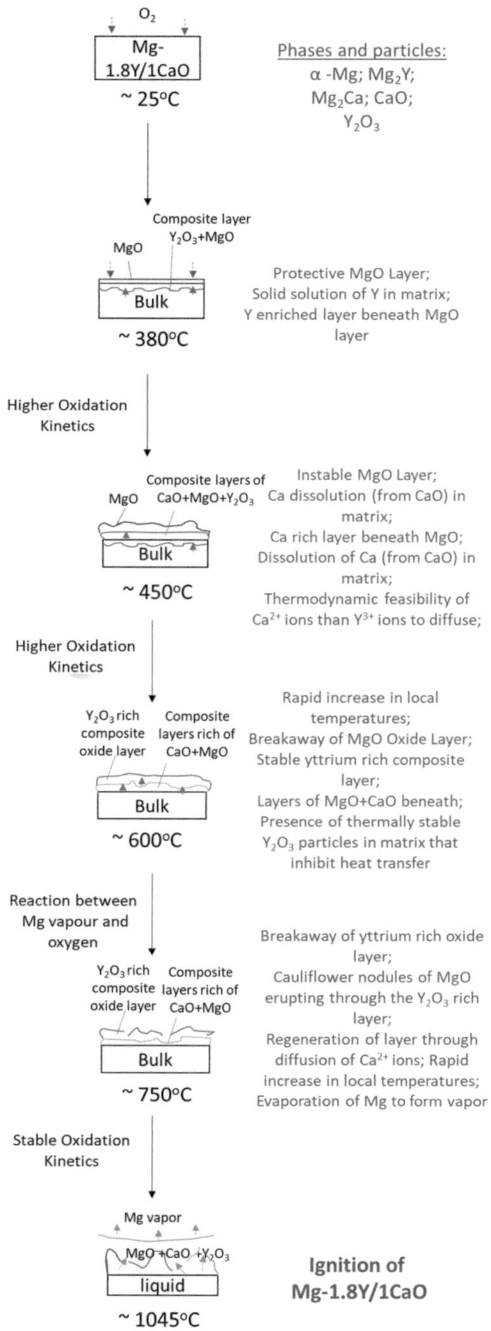

Fig. 6 Mechanism of oxidation and ignition beyond 1000 °C in a Mg−1.8Y/1CaO nanocomposite [2]

strength and ductility at room temperature along with good resistance to corrosion (salt-spray) and superior castability [23]. Hence, they are predominantly being used in the automobile sector as sheets or even engine blocks and other automobile components such as in steering wheels, boot area, etc. For elevated-temperature applications such as engine blocks, newly developed magnesium alloy nanocomposites can be used as they demonstrate good thermal and dimensional stabilities.

Aerospace

With the primary advantage of weight saving, magnesium is a great candidate in aerospace industry to substitute aluminum. Although magnesium was widely applied in the past (1940s) in the commercial airplane construction as a structural material, it was subsequently banned by the Federal Aviation Administration (FAA) to be used due to its perceived easy ignition/flammability. Recently, FAA has lifted the ban and allowed the application of magnesium in the interior of the aircraft. Magnesium renders a weight saving of about 33% against Al and about 77% against steels. A comparison of densities is given in Table 3 along with the distribution of materials that are currently used in commercial aircrafts. Replacement of these heavy and dense metals with magnesium translates to reduction in the consumption of fuel and lowering of the aircraft weight coupled with the reduction in the CO_2 emissions. An aircraft with a greater number of seats has higher amount of weight and therefore greater amount of fuel burnt per passenger. This shows the extremely critical role weight plays in the aerospace sector. Of late, several magnesium alloy composites have been investigated for their ignition temperatures and the enormous improvement observed in the ignition temperatures of magnesium demonstrates the potential of magnesium to be used in the aircraft industry [7]. It is to be noted that Elektron 21 and Elektron WE43 are already FAA approved to be used in the commercial aircraft in cabin applications [2]. A few concerns still persist with respect to the corrosion of magnesium which can be overcome by the adaptation of the coating technology.

Table 3 Density and distribution of materials used in commercial aircraft [24]

Material	Density	% Distribution in commercial aircraft		
		Boeing 777	Boeing 787	Airbus A380
Mg	1.74	–	–	–
Al	2.7	70	20	61
Ti	4.5	7	15	10
Steels	7.9	11	10	–
Composites	1.6–1.9	11	50	22
Others	–	1	5	5

Table 4 Average total shielding effectiveness of magnesium and aluminum [1]

Sample	SE in frequency range of 0.5–6 GHz	SE in frequency range of 7–13 GHz
Mg	65 ± 1	66 ± 4
Al	59 ± 1	65 ± 6

Electronics

In consumer electronic industries, plastic injection molding was revolutionary due to the flexibility it offered in designing casings with complicated designs, colorful appearance and texture. However, the shortcomings such as significantly high EMI emissions and enormous scrap of electronic equipment that is very hard to recycle were realized. Also, it can be added that plastics, generally exhibit poor stiffness and thermal conductivity and bad damping capacity. These attributes are the reason that makes magnesium alloy nanocomposites very lucrative as magnesium can be re-melted, reused and recycled. Its electromagnetic shielding capacity is the same or even superior to that of aluminum alloys as given in Table 4. A key requisite in an electronic equipment that is portable is the ability to withstand elevated ambient/outdoor temperatures and also impact. Magnesium, with excellent stiffness and the best specific damping capacities among known structural materials, is the most viable alternative for these applications. Finally, magnesium also gives a solid looking metallic touch, thereby giving an impression of a good quality and well-graded product [25].

Biomedicine

In human bodies, metals have been used as implants for more than a century, in the form of permanent or non-permanent implants. This is because metals exhibit good mechanical properties (strength, ductility, fatigue resistance and fracture toughness) which are far superior to that of polymers and ceramics. Titanium, steels and Co–Cr alloys are commonly used metallic materials for implant (permanent) applications. Co–Cr alloys were regarded to be toxic and are being avoided currently. Titanium and steels in general have a disadvantage for implant applications despite their excellent mechanical properties as well non-toxicity due to their very high elastic modulus that leads to stress shielding effects between bone and the implant. Magnesium, in this regard, with its elastic/Young's modulus (40–45 GPa) closest to that of the bone, is considered to be one of most suitable lightweight bio-material. With bone being a natural composite, magnesium composites stand a great potential in being suitable in the human body. Further, the elastic matching between the bone and magnesium metal minimizes the stress shielding effect to a great extent. The mechanical properties, i.e. strength and ductility of magnesium alloy nanocomposites can be tailored to exceed that of bone which makes magnesium a mechanically suitable material for

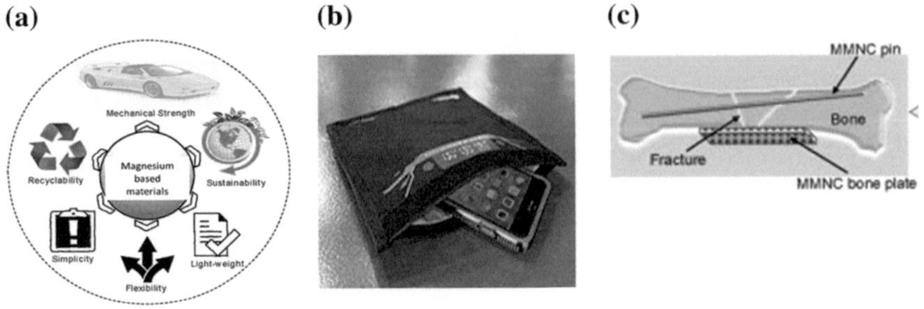

Fig. 7 **a** Advantages of magnesium-based materials in automobile industry; **b** electromagnetic shielding cages that are currently in use in order to prevent electromagnetic pollution, a bag made out of flexible metallic fabric to protect theft of digital data and enhance privacy [27];and **c** uses of magnesium matrix nanocomposites (MMNC) in biomedicine reproduced from [28]

these applications. Human bodies need about 250–400 mg of magnesium/day (it is to be noted that Mg is fourth most occurring ion in human body) and it assists in several functions like aiding bone health, multiple metabolic processes in body besides being anti-bacterial. Since magnesium is both biocompatible and biodegradable, it is the best fit to be used in the body as a non-permanent biodegradable implant as it (i) reduces patient trauma; (ii) requires no revision surgery; (iii) reduces doctor's time; and (d) reduces medical costs. Besides this, the corrosion susceptibilities of magnesium can be used to our advantage. It reacts with simulated biofluids to form localized magnesium ions, increased pH, formation of hydrogen bubbles and a Ca–P layer [26]. With magnesium alloy composites, the reinforcements can be chosen to be biocompatible such that there are no toxicity effects in the body caused due to the reinforcements. Reinforcements like hydroxyapatite, calcium oxide, etc., help in improving the bio-mechanical properties of the material as well as render the material for better biocompatibility. To note here that the corrosion rate of magnesium should be controlled to sync with the healing time of the injury to allow minimal side effects of corrosion, if any. Successful human trials are already made for magnesium-based stents and compression screws while work is in progress for their use as surgical clips and plates. Further efforts need to be made to bring magnesium alloy nanocomposites into the real applications (Fig. 7).

Conclusions

The potential of magnesium alloy-based nanocomposites in terms of their processing, properties and performance are discussed in view of their benefits with respect to light weighting. A snapshot of the diverse and multifaceted capabilities of magnesium-based alloys and composites is provided, which renders magnesium alloy nanocomposites to be potentially dominating metallic materials in widespread engineering and biomedical applications.

References

1. Gupta M (2018) Global emergence and significance of magnesium/technology. Mater Sci Res India 15
2. Tekumalla S, Nandigam Y, Bibhanshu N, Rajashekara S, Yang C, Suwas S, Gupta M (2018) A strong and deformable in-situ magnesium nanocomposite igniting above 1000 °C. Sci Rep 8:7038. https://doi.org/10.1038/s41598-018-25527-0
3. Gupta M (2017) An insight into the capability of composite technology to enable magnesium to spread its wings in engineering and biomedical applications. SOJ Mater Sci Eng 5:1–2
4. Gupta M, Wong WLE (2015) Magnesium-based nanocomposites: lightweight materials of the future. Mater Charact 105:30–46. https://doi.org/10.1016/j.matchar.2015.04.015
5. Tekumalla S, Shabadi R, Yang C, Seetharaman S, Gupta M (2017) Strengthening due to the in-situ evolution of ß1′ Mg–Zn rich phase in a ZnO nanoparticles introduced Mg–Y alloy. Scripta Mater 133:29–32. https://doi.org/10.1016/j.scriptamat.2017.02.019
6. Parande G, Manakari V, Meenashisundaram GK, Gupta M (2017) Enhancing the tensile and ignition response of monolithic magnesium by reinforcing with silica nanoparticulates. J Mater Res 32:2169–2178. https://doi.org/10.1557/jmr.2017.194
7. Tekumalla S, Gupta M (2017) An insight into ignition factors and mechanisms of magnesium based materials: a review. Mater Des 113:84–98. https://doi.org/10.1016/j.matdes.2016.09.103
8. Tekumalla S, Gupta M, Min KH (2018) Using CaO Nanoparticles to improve mechanical and ignition response of Magnesium. Curr Nanomater 3:44–51. https://doi.org/10.2174/2405461503666180502101957
9. Dieringa H (2018) Processing of magnesium-based metal matrix Nanocomposites by ultrasound-assisted particle dispersion: a review. Metals 8, https://doi.org/10.3390/met8060431
10. Chen L-Y, Xu J-Q, Choi H, Pozuelo M, Ma X, Bhowmick S, Yang J-M, Mathaudhu S, Li X-C (2015) Processing and properties of magnesium containing a dense uniform dispersion of nanoparticles. Nature 528:539–543. https://doi.org/10.1038/nature16445, http://www.nature.com/nature/journal/v528/n7583/abs/nature16445.html#supplementary-information
11. Matin MA, Lu L, Gupta M (2001) Investigation of the reactions between boron and titanium compounds with magnesium. Scripta Mater 45:479–486. https://doi.org/10.1016/S1359-6462(01)01059-4
12. Chelliah NM, Singh H, Surappa MK (2017) Microstructural evolution and strengthening behavior in in-situ magnesium matrix composites fabricated by solidification processing. Mater Chem Phys 194:65–76. https://doi.org/10.1016/j.matchemphys.2017.03.025
13. Mounib M, Pavese M, Badini C, Lefebvre W, Dieringa H (2014) Reactivity and microstructure of Al2O3-reinforced magnesium-matrix composites. Adv Mater Sci Eng 2014:6. https://doi.org/10.1155/2014/476079
14. Chen Y, Tekumalla S, Guo YB, Gupta M (2016) Introducing Mg−4Zn−3Gd−1Ca/ZnO nanocomposite with compressive strengths matching/exceeding that of mild steel. Sci Rep 6:32395. https://doi.org/10.1038/srep32395
15. Tekumalla S, Bibhanshu N, Suwas S, Gupta M (2019) Superior ductility in magnesium alloy-based nanocomposites: the crucial role of texture induced by nanoparticles. J Mater Sci 54:8711–8718
16. Tekumalla S, Farhan N, Srivatsan TS, Gupta M (2016) Nano-ZnO particles' effect in improving the mechanical response of Mg−3Al−0.4Ce Alloy. Metals 6:276
17. Availabe online: https://global.kyocera.com/prdct/fc/list/tokusei/bouchou/index.html
18. Czerwinski F (2014) Controlling the ignition and flammability of magnesium for aerospace applications. Corros Sci 86:1–16. https://doi.org/10.1016/j.corsci.2014.04.047
19. Han G, Chen D, Chen G, Huang J (2018) Development of non-flammable high strength extruded Mg–Al–Ca–Mn alloys with high Ca/Al ratio. J Mater Sci Technol 34:2063–2068. https://doi.org/10.1016/j.jmst.2018.03.019
20. Liu M, Shih DS, Parish C, Atrens A (2012) The ignition temperature of Mg alloys WE43, AZ31 and AZ91. Corros Sci 54:139–142. https://doi.org/10.1016/j.corsci.2011.09.004

21. Tekumalla S, Yang C, Seetharaman S, Wong WLE, Goh CS, Shabadi R, Gupta M (2016) Enhancing overall static/dynamic/damping/ignition response of magnesium through the addition of lower amounts (<2%) of yttrium. J Alloy Compd 689:350–358. https://doi.org/10.1016/j.jallcom.2016.07.324
22. Joost WJ, Krajewski PE (2017) Towards magnesium alloys for high-volume automotive applications. Scripta Mater 128:107–112. https://doi.org/10.1016/j.scriptamat.2016.07.035
23. Kulekci MK (2008) Magnesium and its alloys applications in automotive industry. Int J Adv Manuf Technol 39:851–865. https://doi.org/10.1007/s00170-007-1279-2
24. Gupta MGaN (2017) Utilizing magnesium based materials to reduce green house gas emissions in aerospace sectors. Aeronaut Aerosp Open Access J, 1
25. Landkof B (2006) Magnesium applications in aerospace and electronic industries. Magnesium Alloys Appl. https://doi.org/10.1002/3527607552.ch28
26. Gupta M (2018) A snapshot of remarkable potential of mg-based materials as implants. Mater Sci Eng Int J 2:30–33. https://doi.org/10.15406/mseij.2018.02.00030
27. Gupta MGaS (2017) The promise of magnesium based materials in electromagnetic shielding. Jupiter Online J Mater Sci 2. https://doi.org/10.19080/jojms.2017.02.555598
28. Shahin M, Munir K, Wen C, Li Y (2019) Magnesium matrix nanocomposites for orthopedic applications: a review from mechanical, corrosion, and biological perspectives. Acta Biomater 96:1–19. https://doi.org/10.1016/j.actbio.2019.06.007

Nanolayer Surface Phase Change in Self-Healing Materials

Rahul Basu

Abstract The phase change problem in a semi-infinite medium where temperature and concentration are coupled is solved for variable diffusivity. An integral technique is used to solve for concentration and temperature penetration lengths and velocity of change. Surface conditions are held constant whereas diffusivity is allowed to vary slowly corresponding to phase change in the matrix. An analytic solution for variable properties under these conditions is obtained using the Kirchhoff transformation. Penetration lengths are analytically evaluated. A perturbation analysis allows the boundary layer lengths to be estimated, whence the two parameters can be compared. Dimensions obtained by the simulation show that nanomaterial thicknesses are attainable in the boundary layers. Application to phase change concepts relating to self-healing materials is illustrated for porous surface layers.

Keywords Self-healing · Self-healing materials · Composite materials · Surface layers · Porous materials

Nomenclature

a, b	Parameters in the Kirchhoff transformation
c	Concentration
c_p	Specific heat
B	Constant
D	Diffusion coefficient
L, H	Latent heat
s, l	Solid, liquid
n	Power exponent in the approximate scheme
Ste	Stefan number = latent heat/sensible heat

R. Basu (✉)
Department of Mechanical Engineering, Adarsha Institute of Technology, Bangalore 562110, Karnataka, India
e-mail: raulbasu@gmail.com

T. S. Srivatsan and M. Gupta (eds.), *Nanocomposites VI: Nanoscience and Nanotechnology in Advanced Composites*, The Minerals, Metals & Materials Series, https://doi.org/10.1007/978-3-030-35790-0_2

Ti, To	Initial, final temperature
U, V	Transformed temperature, concentration in the Kirchhoff representation
1, 2	Labels for the interacting phases
λ	Eigenvalue for interface velocity
μ	Diffusivity variation for concentration
τ	Time
θ	Temperature (nondim)
θ_m	Fusion temperature
η	Similarity parameter
ε	Recombination or dissolution, small perturbation parameter
Δ	Penetration length (thermal slope is zero)
Δ_1	Penetration length (concentration slope is zero)
δ_m	Thermo-gradient parameter
ζ	Position of the moving phase boundary
λ (τ)	Phase change eigenvalue
α	Thermal diffusivity
α_m	Diffusivity of permeating phase
α_{ij}	Normalized diffusivity $\alpha i/\alpha j$
$\beta(\tau)$	Penetration length for thermal $\Delta = 2\beta\ t^{0.5}$
$\beta_1(t)$	Penetration length for concentration $\Delta_1 = 2\beta_1\ t^{0.5}$
β_0	Separation constant
γ	Temperature-dependent constant for variable diffusivity
Π	3.14159

Introduction

Self-healing materials along with so-called smart materials are now well-known buzzwords. Smart materials appeared earlier utilizing the shape memory effect. By contrast, self-healing materials appeared later after recent conflicts in the new Millenium spurred by Military research into such materials. Traditionally, “healing” in materials involved the use of high temperature to melt the affected part and allow flow or diffusion to augment the repair process, sometimes under pressure. For instance, some plastics can be repaired in this fashion. Thermo-sets typically cannot be softened and remolded for repair, whereas thermoplastics are capable of being softened by heat and restored. This suggests that diffusion of heat and matter are involved. However, there are very few papers dealing with diffusion and transport analyses for material repair. This is probably because practitioners are mainly interested in mechanical and physical properties shadowed by fracture analysis.

Healing has also been studied using the Pennes' bioheat equation for wounds and tissue damage [1] and with the Stefan approach for polymer capacitors, paints, and self-healing polymer composites. Healing of materials may be of the mechanical properties, homogeneity of the matrix, surface texture (as in car paints) and/or

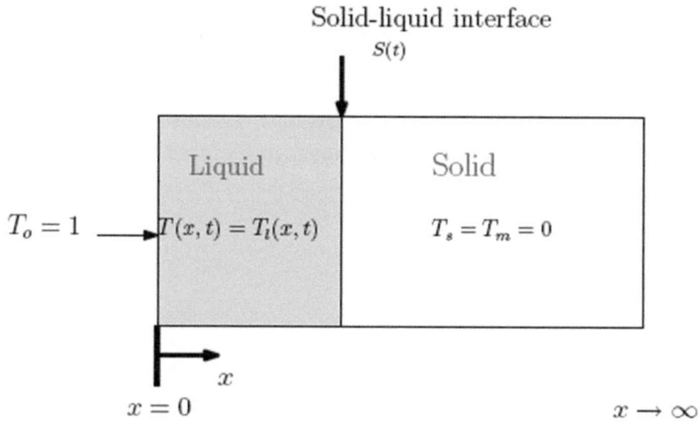

Fig. 1 Schematic of the phase change geometry

other combinations of properties. In the following, the coupling of heat and mass diffusion is explored, leading to a coupled system of partial differential equations. Simplifications allow one to tackle the problem.

There are two main methods by which the repair is carried out: (a) by heat which softens the material above its glass point and allows cracks to rejoin under pressure; (b) by mass diffusion aided chemical reaction and mass diffusion.

Earlier repair of fractured polymers was through plastic welding using the transition temperature of the material. Unfortunately, as in metals, the weld remains the weakest point and constitutes a site for further future damage [2]. For laminate composites, resin injection is often employed to repair damage in the form of delaminating layers. Fiber breakage in laminated composites is treatable by patch reinforcement to partially restore some material strength [3]. None of these methods of repair is a permanent fix; however, they increase the cost by requiring periodic maintenance and service (Fig. 1).

Properties Before and After a Reaction

Due to the mass flux and bonding with subsequent reactions and hardening, the properties of the material after healing will not be the same as before. A time of rest or curing may be needed as in natural healing of damaged tissue, be it animal or plant tissue. The challenge is to formulate and design components that can sense damage and allow in situ healing even during use. It is seen from nature that full repair takes time, however, it has been claimed that for some polymer materials, the repaired mass is as good as or better than the original. A review by Wang [4] describes the capsule and vascular based systems. It is claimed that for capsule systems, the interlaminar fracture toughness can exceed 100% repair. The healing properties of vascular and capsule-based systems are quite different in that vascular systems can repair continuously, whilst capsule systems cannot disperse more fluid after a capsule is ruptured and exhausted. The vascular systems with tubes and capillaries can be refilled whilst the internal capsules cannot. Vascular systems may use inbuilt channels or use sacrificial fibers that are dissolved after manufacture to yield hollow channels, by melting, mechanical removal or change of pH.

Analysis

In this paper, three formulations are examined for rectilinear geometries: Exact solutions for the diffusion equation

(a) With constant coefficients
(b) With variable coefficients
(c) Approximations using integral solutions

The coupled equations are given by the Luikov formulation, [5]

$$\partial\theta/\partial t = \alpha\partial^2\,\theta/\partial x^2 + \varepsilon L/\rho c_p \partial c/\partial t \tag{1}$$

$$\partial c/\partial t = \alpha_m \partial^2\,c/\partial x^2 + \delta_m \alpha_m \partial^2\theta/\partial x^2 \tag{2}$$

In addition, there is a thermal balance across the moving interface, given by (18).

In practice, δ_m is small and one can then decouple the equations using $Z = \theta + \beta_0\,c$, where βo is a decoupling constant, found to be $(1–\alpha/\alpha_m)/\delta_m$.

Subsequently, the decoupled equations in temperature and concentration are considered, more particularly the temperature equation. The concentration equation is amenable to the same treatment for the moving boundary problem but has rarely appeared in the literature.

Boundary Conditions

$$\theta = c = 0 @ x = 0 \tag{3}$$

$$\theta = c = 1 @ x = \infty \tag{4}$$

For the variable properties case, linearly varying conductivity and thermal diffusivity expressions are assumed in the energy and diffusion equations:

$$\partial\theta/\partial t = \partial/\partial x(\alpha_0(1+\gamma\,\theta))\partial\theta/\partial x \tag{5}$$

$$\partial c/\partial t = \partial/\partial x(\alpha_{m0}(1 + \mu c)\partial c/\partial x) + \delta_m \partial\theta/\partial x \tag{6}$$

$$\Delta = 2\beta(\tau)^{0.5},\ \zeta = 2\lambda(\tau)^{0.5} \tag{7}$$

where Δ, Δ_1 are the positions of the temperature and concentration profiles where the slope is zero. ζ is the phase change distance, where $\theta = \theta_m$, $c = c_m$. The procedure for the solution is to first solve for the temperature via the Goodman energy integral technique applied to (6), then applying the temperature solution for the exact case, following the same technique. Setting up the boundary conditions by using the values Δ and ζ as above

Boundary Conditions

$$\theta_2(\Delta, \tau) = 1\ \tau > 0\ x = \Delta(\tau) \tag{8}$$

$$c(\beta, \tau) = 1\ \ x = \beta(\tau) \tag{9}$$

$$\partial\theta_2/\partial x = 0\ \ \tau > 0\ \ x = \Delta(\tau) \tag{10}$$

$$\partial c/\partial x = 0,\ \ x = \beta(\tau) \tag{11}$$

(***Note***: This is an approximation for ease of computation since physically this imposes an adiabatic condition.)

Now integrating Eq. (5) w.r.t. x from $x = \zeta$ to $x = \Delta$ (i.e. from position of phase front to position where slope goes to zero), via the Leibniz rule [6],

$$-\alpha(1 + \gamma\theta)d\theta/dx]_{\zeta}^{\Delta} + \partial\Delta/\partial\tau - \theta_S\delta\zeta/\delta\tau = \partial/\partial\tau \int_{\zeta}^{\Delta}\theta dx \tag{12}$$

Assuming a power law for the thermal and concentration profiles

$$\text{Let } \theta = 1 - (1 - \theta_m)[(\Delta - x)/(\Delta - \zeta)]^n \tag{13}$$

similarly let

$$c = 1 - (1 - c_m)[((\Delta_1 - x)/(\Delta_1 - \zeta)]^n \tag{14}$$

In the references quoted, the exponent n has been taken as >= 2, in the present simulation, it is also allowed to vary for all positive for n <= 2. Substitution of the abovc relations in the integral form (12) and using (7) and canceling out the $(t)^{0.5}$ terms, from Leibnitz' rule

$$(1 + \gamma\theta_m)(1 - \theta_m)/(2(\beta - \lambda)) + \lambda\theta_m = 1 + (1 - \theta_m)(\beta - \lambda)/(n + 1) \tag{15}$$

Collecting terms and simplifying leads to a quadratic equation in (β−λ), the positive root given as

$$(\beta - \lambda) = (n+1)/2\left[-\lambda/2 + \left\{\lambda^2 + 2\alpha n(1+\gamma\theta_m)/(n+1)\right\}^{0.5}\right] \quad (16)$$

Now, for the solid phase, the exact solution for the semi-infinite problem is taken as in Carslaw Jaeger [7], since the conductivity and thermal diffusivity in the solid is much higher than in the liquid, the variation in thermal properties in the expressions above can be neglected, thereby

$$\theta_1 = \theta m \ \mathrm{erf}(x/2(\tau))^{0.5}/\mathrm{erf}(\lambda) \quad (17)$$

Solving for λ, by energy balance

$$\partial\theta_1/\partial \mathrm{x} = \mathrm{k}_{12}\partial\theta_2/\partial \mathrm{x} + \mathrm{H/cp(Ti - To)}\partial\zeta/\partial \mathrm{t} \quad (18)$$

Substituting for θ_1 θ_2 and ζ, simplification gives

$$\theta_m * \exp(-\lambda^2)/\left(\pi^{0.5}\mathrm{erf}(\lambda)\right) + n\ \mathrm{K}_{12}(1-\theta_m)/\left\{(n+1)\left[-\lambda + \left\{\lambda^2 + 2\alpha n(1+\gamma\theta_m)\right\}^{0.5}\right]\right\} = \mathrm{Ste}\ \lambda \quad (19)$$

Results

The following parametric values are used α = 3.3, K = 0.5, γ = 0.1, Ste = 1, θm = 0.95. It is then easy to compute β, as λ is found from Eq. (16) and the profile (13) (Fig. 2 and Table 1).

Following an analogous procedure for the concentration equations, with thermo-gradient coupling taken as zero,

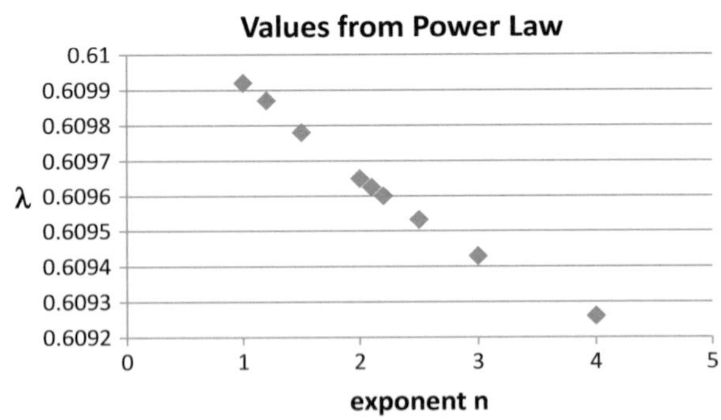

Fig. 2 Lamda (λ) versus exponent 'n' (from Eq. 19)

Table 1 Corresponding Values of beta and lambda

Lambda	Beta-lambda
0.1	0.03882
0.12	0.038465
0.15	0.037958
0.17	0.03765
0.2	0.03723

$$(\beta_1 - \lambda) = (n+1)/2[-\lambda/2 + \{\lambda^2 + 2\alpha\, n(1+\mu\, c_m)/(n+1)\}^{0.5}] \tag{20}$$

where β_1 is the boundary layer (penetration distance)

$$\Delta_1 = 2\beta_1(\tau)^{0.5} \tag{21}$$

Broadbridge et al. [8] have given exact solutions for certain functional dependences of the variable diffusion coefficient. They mention that the reciprocal squared linear diffusivity function transforms into the linear differential equation and has been solved for several types of boundary conditions. A class of solutions with formal analytic solutions where the diffusion coefficient is inversely dependent on concentration was described by Fujita [9].

Exact Solution for the Variable Property Diffusion Equation

Consider the diffusion equation with variable diffusivity:

$$\partial\theta/\partial t = \partial/\partial x[\alpha_0(1+\gamma\theta)\partial\theta/\partial x] \tag{22}$$

In similarity form, setting $\alpha_0 = 1$, this transforms to

$$(1+\gamma\theta)\theta'' + \gamma(\theta')^2 + \eta\theta'/2 = 0 \tag{23}$$

The solution of (23) by Wolfram [10] is attached in Appendix 1.

One way to make this amenable to a boundary layer type of analysis is to look at the damping term (first-order derivative). Normalizing the equation by dividing throughout by the coefficient of the second-order term, it is found to have the coefficient

$$\eta/2[1+\gamma\theta] \tag{24}$$

if γ is small we may take the liberty of replacing θ with the solution to the constant property equation, with the same boundary conditions, which is known to be

$$\theta_m \text{erf x/2 t}^{0.5}/\text{erf}(\lambda) \tag{25}$$

The damping coefficient is then of the form

$$\eta/2[1 + A\ \text{erf}(\eta)], \tag{26}$$

where

$$A = \gamma\theta_m/\text{erf}(\lambda) \tag{27}$$

Converting the equation to standard normal form (by removing the first derivatives) the transformed solution is premultiplied by

$$\exp\{-1/2\gamma \int \eta/2d\eta/(1 + \text{B erf}(\eta)), \text{ where B} = \gamma\theta_m/\text{erf}(\lambda) \tag{28}$$

The boundary layer length may then be defined as that length where the integral evaluates to 2, such that the pre-multiplier tends to 1/e.

Simplification may be attempted where the diffusion coefficient can be replaced by the known solution to the exact constant properties diffusion equation, i.e., $(1 + \gamma\theta) = (1 + \gamma\ \text{erf}(\eta)/\text{erf}(\lambda))$, with $\lambda = 0.25$ (Fig. 3).

The solution obtained by Wolfram [10] shows the following characteristics:

$$(1 + 0.1\ \text{erf(x)})y''(x) + 0.1y'(x)^2 + 0.5xy'(x) = 0 \tag{29}$$

Consider the simplified equation with a known solution in the pre-multiplier term for the variable coefficient. Transforming this equation to standard form with the damping term removed, one gets the pre-multiplier for the second-order term

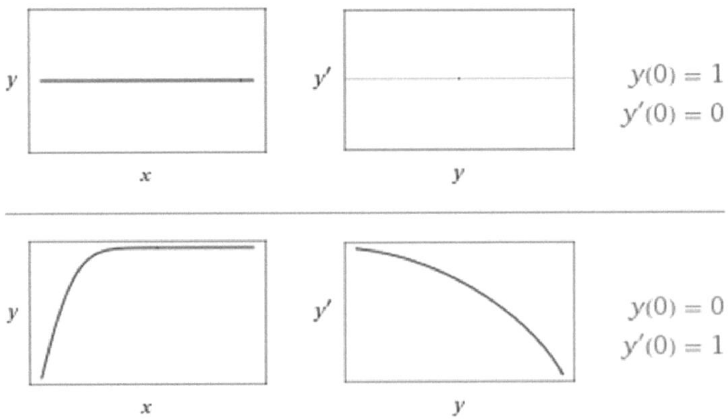

Fig. 3 Plot for the variable diffusivity (solution from Eq. 29)

$$[1 + \gamma\, \theta m \text{ erf}(\eta)/\text{erf}(\lambda)] \quad (30)$$

Hence, the damping term can be found by integration.
The integral has the form $0.5\eta/[1 + \text{B erf}(\eta)]$, where

$$\text{B} = \gamma\ \theta m/\text{erf}(\lambda) \quad (31)$$

Assuming $\text{B} < 1$, this can be expanded as (to first order)

$$0.5\ \eta[1 - \text{B erf}(\eta)] \quad (32)$$

Then the pre-multiplier is $\exp -1/2\ [\ \eta^2/4\text{–B} \int \eta\text{erf}\ (\eta)\ d\eta]$.
From Abramowitz [11], the integral of erf(x) is

$$\text{x erf(x)} - \exp(-x^2)/\pi^{0.5} + \text{const} \quad (33)$$

and from Wolfram [11]

$$\int x \text{ erf}(x)\text{dx} = \frac{1}{4}\left((2\text{x}^2 - 1)\text{erf}(x) + \frac{2e^{x^2}x}{\sqrt{\pi}}\right) + \text{constant}$$

{Integration by parts gives $x^2/2 \text{ erf x} + x/2\pi^{1/2} \exp(-x^2/2) - \text{erf}(x)/2$} (33a)

Evaluation of the pre-multiplier term gives

$$\Delta^2/4 - \gamma\ \theta_m/\text{erf}(\lambda)[0.5\Delta^2 - 0.25 \text{ erf}(\Delta) + 0.5\Delta\ \exp(-\Delta^2)/\pi^{0.5}] = 2 \quad (34)$$

Taking typical values of $\gamma = 0.1$, $\lambda = .0.25$, $\theta_m = 1$, lengths are obtained as in Table 2 (Fig. 4).

Table 2 Attenuation distance versus γ

Distance Δ	γ
2.85381	0.01
2.87995	0.02
2.90689	0.03
2.93467	0.04
2.96332	0.05
3.12157	0.1
3.54196	0.2
4.21694	0.3
5.57467	0.4

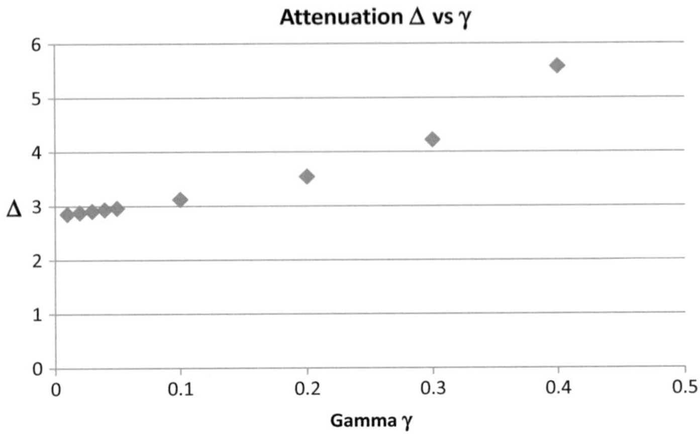

Fig. 4 Plots showing attenuation distance Δ versus γ (from Table 2)

Perturbation Analysis

According to Crank [12], very few perturbation analyses are available for the moving boundary problem. In what follows, the method is outlined up to the second order.

$$\text{Consider the equation } (1+\gamma\ \theta)\theta'' + \gamma[\theta']^2 + \eta\ \theta/2 = 0 \tag{35}$$

Taking the small parameter γ, an expansion is attempted as

$$\theta = \theta_0 + \gamma\ \theta_1 + \gamma^2\theta_2 + \cdots \tag{36}$$

Expanding and collecting terms in γ

$$\theta_0'' + \eta/2\theta' = 0 \tag{37}$$

$$\theta_1 = \text{solution of } \theta_1 + \eta/2\theta_1' = [\theta_o\theta_o'' - \theta_0'^2] \tag{37a}$$

The solutions of Eq. (37) are error functions.

Hence, the first-order solution of the given nonlinear diffusion equation is (from WOLFRAM [10])

$$\theta_0 = \pi^{0.5}\text{Const erf}(\eta/2) + \text{Constant} \tag{38}$$

$$\text{From the B.C. } \theta = 0@\ \eta = 0,\ \text{Constant} = 0 \tag{39}$$

The exponential damping is as before

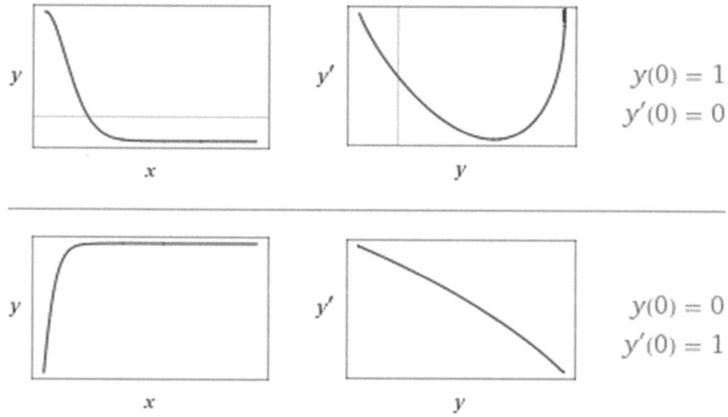

Fig. 5 A solution to Eq. (41)

$$\exp -1/2 \int \gamma/(1+\gamma\ \theta)d\eta \tag{40}$$

where θ is assumed for ease of computation.

The second-order solution involves the solution of the error function equation with an RHS, thus

$$\theta_2'' + \eta/2\theta_2' = \left[\theta_o\theta_o'' - \theta_0'^2\right] \tag{41}$$

where θ_1' involves the first derivative of the error function (Fig. 5).

Analytical Solutions by the Kirchhoff Transformation

In the particular case of the thermal property variation, transformations termed the Kirchhoff transform Ozisik, [6] can be applied. Given the thermal and mass balance equations as before with variable thermal and mass diffusive coefficients, a substitution is made

$$\mathrm{U} = \alpha(\theta')/\alpha_0 \mathrm{d}\theta',\ \mathrm{V} = \beta(c')/\beta_0 \mathrm{d}c' \tag{42}$$

The transformed equations become

$$\partial\mathrm{U}/\partial\mathrm{t} = \alpha\partial^2\mathrm{U}/\partial\mathrm{x}^2 \tag{43}$$

$$\partial\mathrm{V}/\partial\mathrm{t} = \alpha_m\partial^2\mathrm{V}/\partial x^2 + \gamma\ \delta m\ \partial^2\mathrm{U}\partial/x^2\ (\gamma = a_0\ a_{m2}/a\ a_{m0}) \tag{44}$$

The form of the solution to Eqs. (436) and (44) is known from an earlier solution to constant properties. Hence, the same solutions are used with $\alpha = \alpha(\theta)$, $\alpha_m = \alpha_m(c)$. Since the equations are in U, V, x and t and the diffusion constants are independent of x, and t, but depend on T and c, known solutions can be used for the transformed diffusion equations. The separation techniques used earlier can be applied if slowly varying coefficients are assumed, i.e. dz/dt and da/dx = 0. Assuming a separation constant β_0, where $Z = U + \beta_0 V$, $\beta_0 = (1-\alpha/\alpha m_2)(\alpha\ \alpha m_0)/(\alpha_0\ \delta\ \alpha\ m_2)$.

In the following, quadratic variations are assumed, i.e.,

$$U = T + a\,T^2/2,\ \ V = c + b\,c^2/2 \tag{45}$$

For the interfacial balance, the following transforms are applied:

$$\partial U/\partial T = \alpha(T)/\alpha_0,\ \ \partial T/\partial x = \partial T/\partial U/\partial U/\partial x \tag{46}$$

$$K(T)\partial T/\partial x = K\,\alpha_0/\alpha(T)\partial U/\partial x \tag{47}$$

Differentiating the known solutions in U and V for the solid and moisture regions, the thermal balance at the interface can be written as before. A transcendental equation has to be solved again for the interface location. The analysis is capable of being extended to the case where the interfacial temperature is unknown. In the present analysis, this interfacial temperature is taken at a constant value.

To analyze the transformed equation, consider a similarity transformation $\eta = \mathbf{x}/(\mathbf{t})^5$, which transforms t to the form

$$-\eta/2\ \partial U/\partial\eta = \alpha'\partial^2 U/\partial\eta^2 \tag{48}$$

Consider the second term (damping), and the transformation to remove the second term which is exp-1/2 $[\eta^2/4\ \alpha]$.

The boundary layer defined as before is

$$[8\ \alpha]^{0.5} \tag{49}$$

Now since $U = T + aT^2/2$ by the Kirchhoff transformation, T can be obtained by solving for the roots of the quadratic, (48)

Substituting this in the transformed equation, and neglecting the product T dT/dη, the damping terms are seen to be

$$\partial^2 T/\partial\eta^2 + \partial T/\partial\eta[\boldsymbol{+\eta/2\alpha + (\alpha'/2\alpha)T_0}] = 0, \tag{50}$$

Assume $T = T_0 + \varepsilon\ \eta + -$, then the damping term is then seen to be modified as the integral of the highlighted portion giving

$$+\eta^2/4a + (\alpha'/\alpha)T_0\eta \tag{51}$$

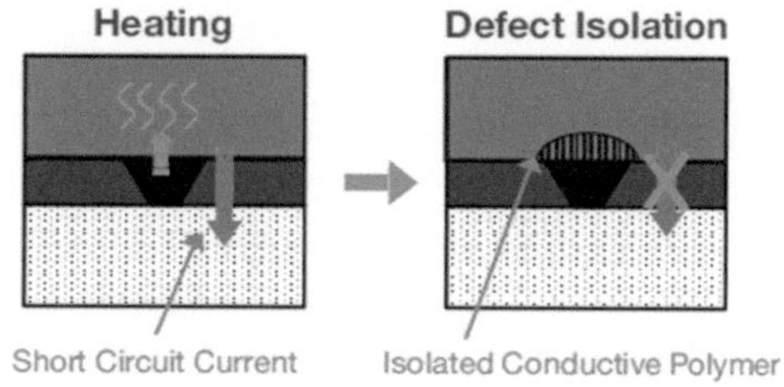

Fig. 6 Schematic of polymer self healing in a capacitor (from [15])

Equating (51) to 2 to give damping of 1/e, the length can be solved for and one obtains:

$$-2\alpha' + / - \mathrm{sqrt}[8\alpha + 4\mathrm{T}_0^2(\alpha')^2] \qquad (52)$$

Hence, depending on the functional form of α with T, the length differs from that in (49) obtained for the transformed variable U, whilst (52) gives the analogous length for T (or θ).

The parameter of interest for the nano range is **area/mass**. Hence, the smaller the mass per unit area, the smaller the ratio gets and one can assert that for a film this is attainable. Soumya [13] illustrates diffusivity of the order 10^{-7} m^2/s. From (49), the boundary layer is of the order of 10^{-3} m to 10^{-4} m. The volume of the boundary layer per area will vary as the layer thickness. For Carbon Aerogel, (nanometer size carbon particles with porous structure), a ratio of 10^{-3} was measured [14], which is consistent with the estimate for the ratio in the boundary layer above (Fig. 6).

Commercial Applications

Specific studies and description of commercially available self-healing composite products are difficult to find in the open literature. Some reviews have appeared describing self-healing polymers [16–19]. In contrast, there are several descriptions of concrete with repair and self-healing mechanisms, where time factors have been estimated for repair, as in [20]. A discussion of permeability and Darcy flow in fiber-reinforced cement is available in [21]. Capacitors with the repair of polymer insulator after high voltage breakdown have been designed by Panasonic Co.

Discussion

It is seen that there are two lengths (Δ and λ) differing by an order of magnitude. The first is due to thermal conductivities and the second due to the moving boundary which defines the transformation length, working with a similarity parameter combining distance and time. Using the definition of the boundary layer as that length where the attenuation goes as 1/e, the first is closely connected to the Fourier number (appearing as the coefficient of the damping term in the form of the similarity variable), and the second related to the Stefan number (physical motion or transport of heat by the moving boundary). It is found from the exact and approximate solutions that a first-order approximation can be realized for the moving boundary portion. The attenuation of the Fourier component shows various numbers which are higher than the Moving Boundary component. The higher order approximations may no doubt give closer correspondence; however, in the interest of space and time, these have been left for further analyses. The time for permeation and diffusion especially with a moving boundary can be estimated, at least to first order for engineering purposes. In the area of wound healing, similar concepts with growth and dissolution can be applied to the same type of equations using concentration since latent heat of transformation may not be applicable [21–23]. For the concentration solution, similar calculations apply with the inclusion of the thermo-gradient parameter. Actual verification of these boundary length estimates may require microanalysis by EDX or other methods. Verifications have appeared only for diffusivities by experimental or inverse methods, [24–26]. Figures 7 and 8 give solutions for variable diffusion properties.

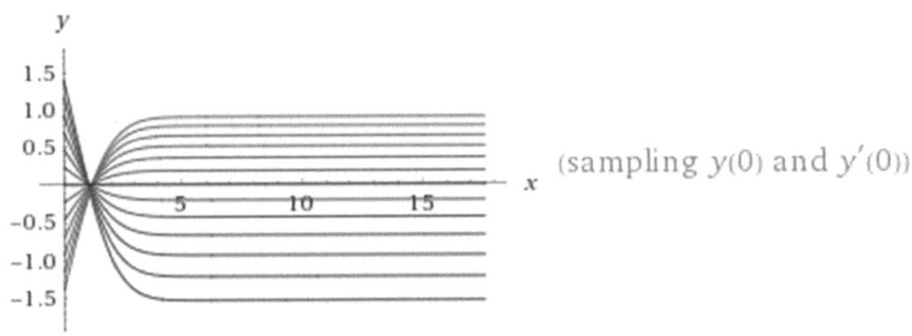

Fig. 7 The second-order solution given by Eq. (43) [with RHS a constant]

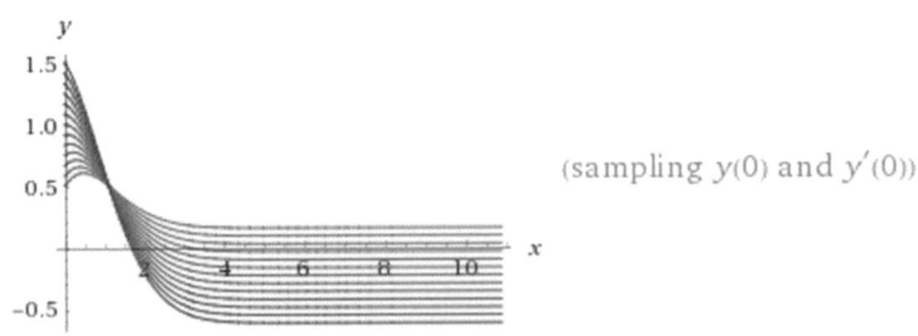

Fig. 8 Contour plot of the solution of Eq. (43) near the origin

Conclusions

1. The analysis and modeling of various self-healing mechanisms rely on inherent properties of heat and mass flow down a gradient in porous media, which automatically directs the flow of heat and liquid to the part with vacancies.
2. In the case of solid-state diffusion, the mechanism depends on the concentration and thermal gradient and is described by mass and thermal diffusion mechanisms.
3. The time of availability of the repair materials depends on various inherent properties of the matrix, and it is possible to calculate the quantitative effects, and thus anticipate the length and time for repair and transformation.
4. Depending on the thickness of the transformation layer, one may engineer nanoparticles for various applications. One may increase the surface area to volume ratio by increasing the number of particles, reducing the radii and/or introducing porosity.
5. A design methodology for such self-repairing materials may thus be formulated using these parameters and models.

References

1. Gupta N, Shakra M (2016) A two-dimensional mathematical model to analyze thermal variations in skin and subcutaneous tissue region of human limb during surgical wound healing. Appl Math 7:145–158
2. Liu D, Lee CY, Lu X (1993) Reparability of impact-induced damage in SMC composites. J Compos Mater 27(13):1257–1271
3. Osswald T, Menges G., (2003) Failure and damage of polymers. In: Osswald T, Menges G (eds) Materials science of polymers for engineers, p 447, Hanser Publishers, Munich, Germany
4. Wang Y, Pham Duc, Ji Chiungian (2015) Self healing composites: preview. Cogent Eng 2:1. https://doi.org/10.1080/23311916.2015.1075686
5. Luikov AS (1975) Systems of differential equations of heat and mass transfer in capillary-porous bodies (review). Int J Heat Mass Transf 18(1):1–14
6. Ozisik MN (1980) Heat conduction. Wiley, New York
7. Carslaw HS, Jaeger JC (1959) Conduction of heat in solids. OUP, Oxford, UK
8. Broadbridge P, Tritscher P, Avagliano A (1993) Free boundary problems with nonlinear diffusion. Math Comput Model 18(10):15–249
9. Fujita H (1952) The exact pattern of concentration-dependent diffusion in a semi-infinite medium. Textile Research J 22:757–760
10. www.wolframalpha.com. Accessed Feb 2018
11. Abramowitz M, Stegun I, (eds) (1972) Handbook of mathematical functions. Dover, New York
12. Crank J (1985) Free and moving boundary problems. Oxford University Press, Oxford, UK
13. Soumya H, Philip J (2016) Thermal diffusion in thin plates and coatings: influence of thickness and coating material. Turkish J Phys 40:256–263
14. Biener J, Wittstock J, Baumann T et al (2009) Surface chemistry in nano scale materials. Materials 2:2404–2428. https://doi.org/10.3390/ma2042404
15. www.panasonic.com/industrial. Accessed Jan 2018
16. Wang Y (2014) Dynamic urea bond design of reversible self healing polymers. https://doi.org/10.1038/ncommms4218

17. Mphahlele K, Roy SS, Kolesnikov A (2017) Self healing polymeric composite material design, failure design, and future outlook: a review. Polymers 9:537. https://doi.org/10.3390/polym9100535
18. Nakamura M, Takeo K, Usuda T, Ozaki S (2017) Finite element analysis of Self Healing and Damage Processes in Alumina/SiC Composite Ceramics. Technologies 5:40. https://doi.org/10.3390/technologies5030040
19. Privman V, Dementsov A, Sokolov I (2018) Modeling of self-healing polymer composites reinforced with Nanoporous glass fibers. www.clarkson.edu/Privman. Accessed Jan 2018
20. Formia A, Irico S, Bertola F et al (2016) Experimental analysis of Self Healing cement-based materials incorporating extruded Cementitious hollow tubes. J Intell Mater Syst Struct 27(9):2633–2652
21. Homma D, Mitashi H, Nishikawa T (2009) Self healing capacity of fiber reinforced cementitious composites. J Adanced Concrete Technol 7(2):217–228
22. Trask RS, Williams GJ, Bond IP (2007) Bioinspired self-healing of advanced composite structures using hollow glass fibers. J Royal Soc Interf 4(13):363–371
23. Strickland S, Hin M, Sayanagi MR et al (2014) Self healing dynamics of surfactant coating on thin viscous films.arxiv:1309.3537v2
24. Piserchia A, Barone V (2016) Towards a general yet effective computational approach for diffusive problems. J Chem Theory Comput 12(8):3482–3490
25. Pavlov TR, Staicu D, Vlahovic L et al (2018) A new method for the characterization of temperature-dependent thermophysical properties. Int J Therm Sci 124:98–109
26. Fernandes AP, Sousa PFB, Borges VL, Guzmaras G (2010) Use of 3D transient analytical solutions based on Greens Functions to reduce computational time in inverse conduction problems. Appl Math Model 34:4040–4045

Part II
Microstructure and Properties of Nanocomposites

On the Role of Processing on Microstructural Development and Mechanical Response of Magnesium-Based Nanocomposites

S. Sankaranarayanan, S. Jayalakshmi, Arvind Singh, T. S. Srivatsan and Manoj Gupta

Abstract In this paper, an attempt is made to present and briefly elucidate and highlight the intricacies specific to magnesium-based composites containing nanosized particulates of boron nitride [BN]. The composite materials were engineered using the techniques of solid-state processing and liquid-state processing. The solid-state processing method was based on use of the powder metallurgy (PM) approach, which essentially involved use of the technique of microwave-assisted bidirectional sintering. The liquid-state processing method involved the technique of disintegrated melt deposition [DMD], which brings together the benefits of stir casting, bottom pouring and spray deposition. The composite materials, in the as-synthesized condition, were subsequently extruded. Samples of the as-extruded composites were characterized for their microstructure and basic mechanical properties. The influence of processing technique used on microstructural development and resultant mechanical properties of the engineered composites is presented and briefly discussed.

Keywords Magnesium-based composites · Nanosized reinforcement · Processing · Microstructure · Mechanical properties

Introduction

Lightweight materials are important for energy-efficient and cleaner transportation means. Since it takes less energy to accelerate a lighter vehicle, light-weighting of both automotive components and aerospace components exerts an influence on fuel

S. Sankaranarayanan (✉)
ANSYS Software Pvt. Ltd, Bangalore 560034, Karnataka, India
e-mail: ssnseetharaman@yahoo.com

S. Sankaranarayanan · S. Jayalakshmi · A. Singh · M. Gupta
Department of Mechanical Engineering, National University of Singapore, Singapore, Singapore

S. Jayalakshmi · A. Singh
Wenzhou University, Wenzhou 325035, Zhejiang, China

T. S. Srivatsan
Department of Mechanical Engineering, The University of Akron, Akron, OH 44325, USA

T. S. Srivatsan and M. Gupta (eds.), *Nanocomposites VI: Nanoscience and Nanotechnology in Advanced Composites*, The Minerals, Metals & Materials Series, https://doi.org/10.1007/978-3-030-35790-0_3

consumption for purpose of better fuel efficiency and energy savings [1]. For example, in a V6 cylinder car, replacement of a cast-iron engine block with magnesium (Mg) reduced the weight from 86 to 30 kg, which in turn promoted noticeable savings in fuel consumption and cost [2]. Similarly, BMW also reduced the fuel consumption of its model cars by as much as 30 pct. (from 1990 to 2007) using advanced lightweight engine concept [3].

Of the metallic materials available, magnesium, aluminum and titanium with their relatively low densities (in comparison to steel, nickel and copper.) qualifies for applications requiring a reduction in weight [4]. While the metals and their alloy counterparts have high specific mechanical strength (σ/ρ), their strength properties can further be improved by engineering or synthesizing metal matrix composites, i.e. through the addition of ceramic reinforcements of desired volume fraction to a metal matrix. Extensive research is continuously being conducted in this regard to develop new light metal alloys and their composites [5]. Titanium alloys and composites possess very high tensile strength and toughness (at both high and low temperatures) coupled with a combination of lightweight, extraordinary corrosion resistance, and ability to withstand extreme temperatures. However, the high cost and processing difficulties limits their use to selective critical engineering applications [5]. So far, aluminum alloys and composites have been most successfully manufactured and commercialized for various industrial applications and their behavior have been widely studied and applied to different engineering structures [5, 6].

The magnesium alloys also exhibit mechanical properties that are comparable with aluminum alloys. Hence, the research and development of magnesium-based materials received an exponential interest since magnesium is ~35% lighter than aluminum. It also possesses excellent damping capacity, machinability and castability, and can safely be considered as a prospective candidate material for future applications encompassing a wide range of engineering sectors [7]. While magnesium and its alloys offer greater degree of weight savings, the use of magnesium in its pure form is often restricted in critical industrial and commercial applications due to their poor elastic modulus, strength, toughness and ductility. To overcome these limitations, it is conventionally alloyed with metallic elements such as aluminum (Al), zinc (Zn), zirconium (Zr) and even the rare earth (RE) metals. Despite the development of new alloys, most commercial alloys exhibit low thermal stability and are stable only up to ~150 °C. In this regard, the incorporation of hard and brittle ceramic reinforcements into the magnesium matrix significantly improves the dimensional stability, elastic modulus, strength, hardness and wear resistance. However, the ductility is often compromised. To minimize the loss in both ductility and toughness while concurrently retaining strength at acceptable levels, the addition of nanoscale reinforcements is promising [7, 8]. An efficient dispersion of the nanoscale reinforcements in a magnesium or magnesium alloy metal matrix aids in enhancing dispersion strengthening while concurrently activating non-basal slip for the purpose of achieving superior strength and acceptable ductility.

In the present research study, nanosized boron nitride (BN) particles were incorporated into a pure magnesium (Mg)-matrix using the techniques of solid-state powder metallurgy and the liquid-state disintegrated melt deposition (DMD). Boron nitride

(BN) is an interesting reinforcing material due to the unique combination of properties it offers to include the following: (i) a low density, (ii) high melting point, (iii) high thermal conductivity, and (iv) high electrical resistivity. This makes it an ideal candidate for selection as the reinforcing phase. The developed magnesium/boron nitride [Mg/BN] nanocomposites were then extruded prior to being tested for their physical properties, microstructural development and mechanical properties. The influence of the boron nitride (BN) reinforcing phase and the processing technique used on microstructural development and mechanical properties were established using process-structure-property relationships.

Experimental

Materials

In this present study, while pure elemental magnesium turnings (purity >99.9%) provided by ACROS organics [Morris Plains, NJ, USA] was used as the base metal for the technique of DMD. The magnesium (Mg) powder [of purity: >98.5%] and with a size range 60–300 μm, was provided by Merck (Darmstadt, Germany) and used for the powder metallurgy (PM) method. Similarly, boron nitride (BN) powder of size ~50 nm and purity >99.5% [provided by NaBond (Hong Kong, China)] was used as the reinforcing phase.

Primary Processing

Pure Mg and the nano-BN particulate reinforced composites required for the current study (details as provided in Table 1) were synthesized using the techniques of: (a) solid-state powder metallurgy (PM), and (b) liquid-state disintegrated melt deposition (DMD) and the details of chemical composition is listed in Table 1.

Table 1 Details of materials developed in this study

Powder metallurgy		Disintegrated melt deposition	
Materials	Vol.% of nano-BN	Materials	Vol.% of nano-BN
Pure Mg	0	Pure Mg	0
Mg-0.3 BN	0.3	Mg-0.3 BN	0.3
Mg-0.9 BN	0.9	Mg-0.6 BN	0.6
Mg-1.4 BN	1.4	Mg-1.2 BN	1.2

Blend-Press-Sinter Powder Metallurgy (PM) Method

The powder metallurgy process essentially involved blending pure magnesium powder with the required amount of nanosized boron nitride (BN) powder in a mechanical alloying machine [Model: RETSCH PM-400 (RETSCH, Hanna, Germany)] at 200 rpm for 1 h (60 min). No balls or process control agent was used during the blending step. The blending process was carried out without any protective atmosphere. The blended magnesium (Mg)-powder mixture was subsequently uniaxially cold compacted using a pressure of 50 tons to get billets that measured 35 mm in diameter and 40 mm in length. The compacted billets were then sintered using a hybrid microwave-assisted two-directional sintering technique. The as-provided billets were heated for 14 min to a temperature near the melting point of magnesium (Mg) in a 900 W, 2.45 GHz microwave oven [Model: Sharp] using colloidal graphite as the oxidation barrier layer [9].

Disintegration Melt Deposition (DMD) Method

The technique of disintegrated melt deposition [DMD] involved heating of the magnesium turnings together with the required amount of nano-BN in a graphite crucible in an environment of argon gas up to 800 °C. To facilitate a near-uniform distribution of the reinforcing particulates, the superheated molten slurry was gently stirred at 460 revolutions per minute using a twin-blade (pitch 45°) mild steel impeller (coated with Zirtex 25) for 8 min. The composite melt was subsequently bottom poured into a steel mold followed by disintegration using two jets of argon gas that were oriented normal to the melt stream. Following deposition, an ingot that measured 40 mm in diameter was obtained. The synthesis of monolithic (pure) magnesium was carried out using similar procedure except without the addition of reinforcements. The ingots obtained using the technique of DMD were subsequently machined to get billets that had a diameter of 35 mm and a height of 40 mm [10].

Secondary Processing

For secondary processing, the billets that were prepared using the techniques of powder metallurgy (PM) and disintegrated melt deposition (DMD) were coated with colloidal graphite and soaked at 400 °C for full 60 min and then hot extruded at 350 °C using an extrusion ratio of 20.25:1, to obtain rods that had a diameter of 8 mm. Samples from the extruded rods were used for continued microstructure examination and mechanical property characterization.

Materials Characterization

Density Measurements

Based on Archimedes principle, the experimental densities of pure magnesium (Mg) and Mg/BN nanocomposites fabricated using the techniques of powder metallurgy (PM) and disintegrated melt deposition (DMD) were measured using the extruded rods in the as-polished condition. For determining the porosity, the experimental density values were compared with the theoretical values that were calculated using the theory of rule-of-mixtures.

X-Ray Diffraction Studies

Samples taken from the extruded rods of pure Mg and Mg/BN nanocomposites fabricated using the techniques of powder metallurgy (PM) and disintegrated melt deposition (DMD) were exposed to Cu Kα radiation of wavelength $\lambda = 1.54056$ Å and at a scan speed of 2°/min by using an automated diffractometer [Model: Shimadzu lab-X XRD-6000] (Kyoto, Japan). The Bragg angles and the values of interplanar spacing *(d)* obtained were subsequently matched with the standard values for magnesium (Mg), boron nitride (BN) and related phases. Further, the basal plane orientation of the as-developed magnesium-based materials was analyzed based on the X-ray diffraction (XRD) peaks obtained from: (i) the cross section and (ii) the longitudinal section of the extruded rods.

Microstructural Characterization

Microstructural characterization studies were conducted on metallographically polished extruded samples in conjunction with a field emission scanning electron microscope (FESEM) (Model: Hitachi S-4300 (Tokyo, Japan)), an Olympus metallographic optical microscope (Tokyo, Japan) and Scion image analysis-based software (Sacramento, CA, USA).

Tensile Test and Compressive Test

The smooth bar tensile and compressive properties of pure Mg and Mg/BN nanocomposites were determined at ambient temperature [27° C] as per procedures detailed in the standards ASTM E8-11 and E9-05. The tensile tests were conducted on round tension test specimens having a diameter of 5 mm and a gauge length of 25 mm using a fully automated servo-hydraulic mechanical testing machine [Model: MTS-810]. The crosshead speed was set to 0.254 mm/min. Similarly, the compressive tests were conducted on round test specimens having a diameter of 7 mm and length of 10 mm and the crosshead speed was maintained at 0.04 mm/min. For each composition, five samples were tested to ensure repeatable values.

Table 2 Results of density and porosity measurements

Material	Processing method	Theoretical density (g/cc)	Experimental density (g/cc)	Porosity (%)
Powder metallurgy method processed materials				
Pure Mg	PM	1.7400	1.7365 ± 0.0011	0.20
Mg-0.3BN	PM	1.7436	1.7391 ± 0.0034	0.27
Mg-0.9BN	PM	1.7506	1.7458 ± 0.0014	0.21
Mg-1.4BN	PM	1.7579	1.7512 ± 0.0019	0.38
Disintegrated melt deposition method processed materials				
Pure Mg	DMD	1.7400	1.7380 ± 0.0006	0.11
Mg-0.3BN	DMD	1.7436	1.7412 ± 0.0010	0.14
Mg-0.6BN	DMD	1.7474	1.7446 ± 0.0013	0.16
Mg-1.2BN	DMD	1.7548	1.7515 ± 0.0009	0.19

Results and Discussion

Density and Porosity

The experimental density of magnesium (Mg) materials measured utilizing Archimedes principle (Table 2) was found to accord reasonably well with the calculated theoretical density. With the addition of nano-BN particulate reinforcements to the magnesium metal matrix, only a marginal increase in the density values of pure magnesium was observed. The porosity value of pure magnesium was found to increase with the addition of nano-BN reinforcements. On relative scale, the powder metallurgy (solid phase) processed materials were found to possess more porosity when compared to the disintegrated melt deposition (DMD) processed materials.

X-Ray Diffraction

Figure 1 shows the X-ray diffractograms of pure magnesium (Mg) and a representative Mg/BN nanocomposite sample that were synthesized using the techniques of PM and DMD and obtained from both the transverse (T) and longitudinal (L) section of the hot extruded samples. While the high-intensity magnesium peaks were prominently seen, the peaks corresponding to boron nitride (BN) were not easily visible in extruded Mg/BN nanocomposites. This is essentially due to their relatively low volume fraction (<2.5 vol.%) in the magnesium matrix, which remains undetected by the technique of x-ray diffraction (XRD). However, the presence of these reinforcements can be confirmed from a careful examination of the microstructure (details provided in the following section).

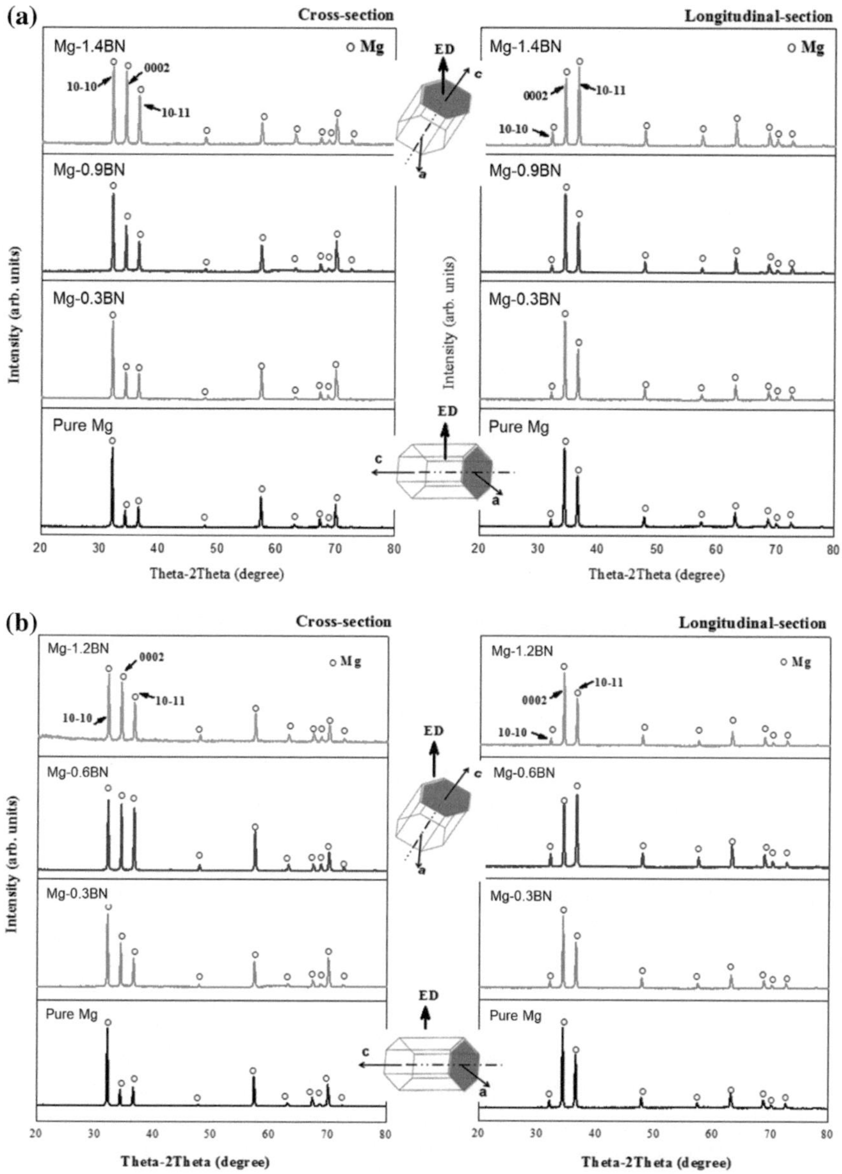

Fig. 1 X-ray diffractograms of **a** PM and **b** DMD processed pure Mg and Mg/BN nanocomposites

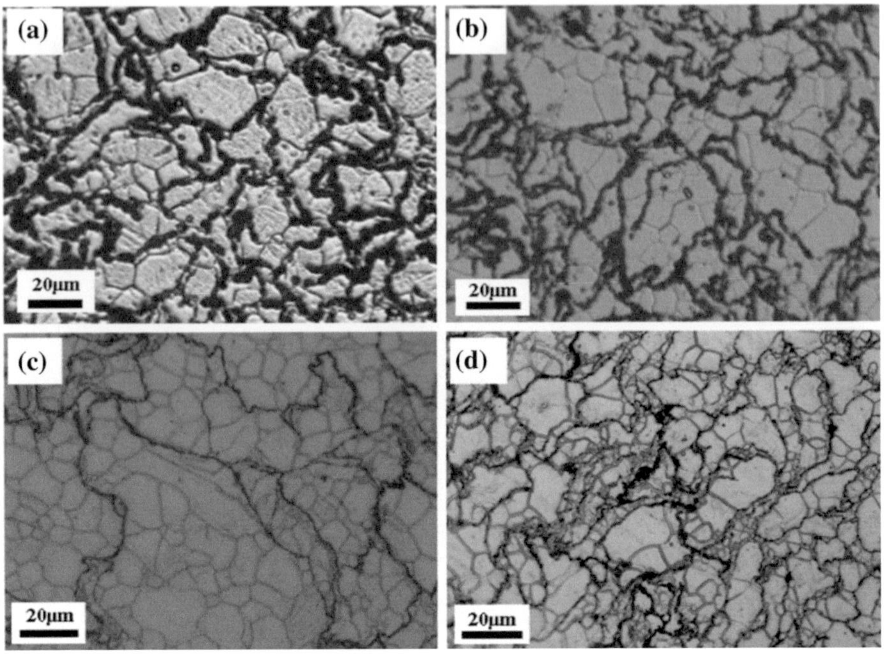

Fig. 2 Optical micrographs showing the grain characteristics of PM processed **a** pure Mg and **b** Mg-0.3BN, **c** Mg-0.9BN and **d** Mg-1.4BN

The results of X-ray analysis were used to interpret the effect of nano-BN addition on the crystallographic orientation (specifically, the basal plane orientation) of the Mg-matrix, since it is well known, that presence of the reinforcing phases does contribute to a change in basal orientation (texture) of the Mg-crystal [11, 12]. The peaks observed at $2\theta = 32°$, $34°$ and $36°$ diffraction pattern of the developed as-extruded Mg materials correspond to the (1 0 − 1 0) prism, (0 0 0 2) basal and (1 0–1 1) pyramidal planes of the HCP Mg-crystal. Along the cross section, the prismatic plane intensity ($2\theta = 32°$) was maximum in all the cases. This clearly indicates that any of the prismatic planes are aligned in a direction, which is perpendicular to the extrusion direction. Also, the basal plane intensity ($2\theta = 34°$) was found to increase following the addition of nano-boron nitride (BN) particulate reinforcements. This provides evidence that most of the basal planes are aligned parallel to the extrusion direction, indicating a strong basal texture, which is commonly observed in the wrought magnesium materials [9].

Microstructure

The results of grain size measurements conducted on the optical micrographs (Figs. 2 and 3) of the powder metallurgy (PM) processed and disintegrated melt deposition

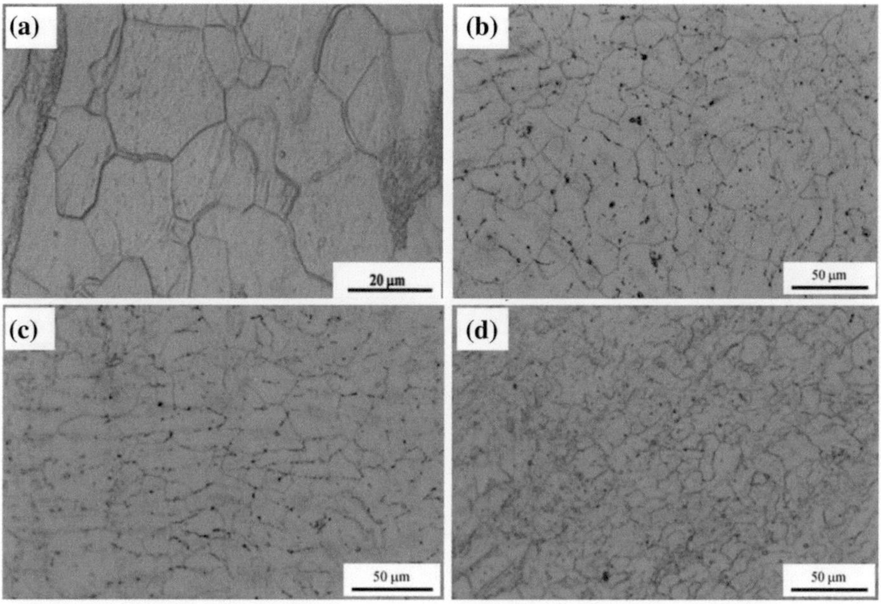

Fig. 3 Optical micrographs showing the grain characteristics of DMD processed **a** pure Mg and **b** Mg-0.3BN, **c** Mg-0.6BN and **d** Mg-1.2BN

(DMD) synthesized pure magnesium (Mg) and magnesium/boron nitride (Mg/BN) nanocomposites are shown in Table 3. It revealed a unimodal distribution of grain size, which indicates one maximum for the grain size distribution, i.e., maximum grains of the same size. This reveals the occurrence of complete recrystallization of the Mg-matrix during extrusion [13]. Although the average grain size of both the DMD and PM synthesized hot extruded pure Mg decreased with the addition of boron nitride (BN) nanoparticles, the difference is only marginal considering the standard deviation.

Table 3 Results of grain size measurements

Material	Processing method	Grain size (μm)	Aspect ratio
Pure Mg	PM	29 ± 6	1.61 ± 0.55
Mg-0.3BN	PM	21 ± 3	1.54 ± 0.44
Mg-0.9BN	PM	22 ± 1	1.67 ± 0.31
Mg-1.4BN	PM	19 ± 4	1.56 ± 0.39
Pure Mg	DMD	28 ± 11	1.82 ± 0.73
Mg-0.3BN	DMD	18 ± 6	1.76 ± 0.49
Mg-0.6BN	DMD	15 ± 6	1.63 ± 0.42
Mg-1.2BN	DMD	9 ± 8	1.61 ± 0.37

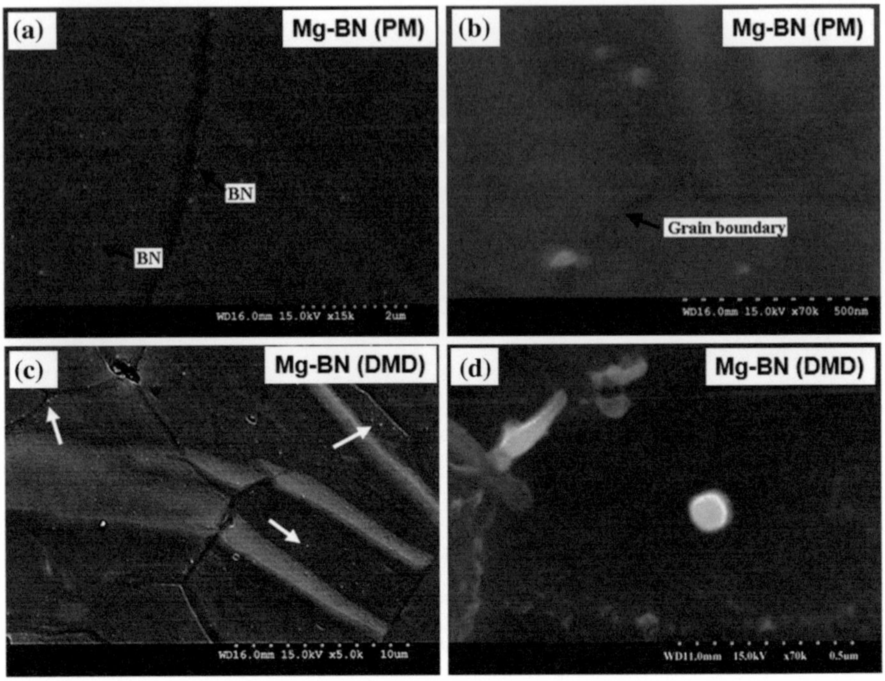

Fig. 4 Representative SEM micrographs showing the distribution and interfacial characteristics in PM and DMD processed Mg/BN nanocomposites

The results of microstructural characterization studies (Fig. 4) indicate a near-uniform distribution of the nano-BN particulates both within the Mg-matrix and near the grain boundaries of both the powder metallurgy (PM) and disintegrated melt deposition (DMD) processed Mg/BN nanocomposites. This is essentially attributed to the following two reasons:

(i) A close density value between the Mg-matrix (1.7 g/cc) and the BN reinforcements (2.98 g/cc), that is favourable for the occurrence of less gravity-assisted segregation problems during primary processing.
(ii) Suitable selection of both blending and stirring parameters used during PM and DMD methods, and
(iii) An overall efficient extrusion process that can break down the agglomerates (if any present and dispersed through the microstructure). Literature study reveals that a homogeneous distribution of the reinforcing particulates can be achieved in the case of extruded materials for a large deformation strain (extrusion ratio–22.5:1), regardless of the size difference between the matrix and the reinforcing particulates [14].

Table 4 Room temperature mechanical properties of PM and DMD processed pure Mg and Mg/BN nanocomposites

S. no.	Material	Tensile properties			Compressive properties		
		0.2 YS [MPa]	UTS [MPa]	Fracture strain [%]	0.2 CYS [MPa]	UCS [MPa]	CFS [%]
Powder metallurgy method processed materials							
1	Pure Mg	136 ± 8	170 ± 7	6.1 ± 1.2	70 ± 2	250 ± 7	24.5 ± 2.7
2	Mg-0.3BN	127 ± 6	192 ± 8	8.8 ± 1.9	88 ± 6	290 ± 9	20.9 ± 1.8
3	Mg-0.9BN	142 ± 4	200 ± 5	8.6 ± 0.5	108 ± 2	312 ± 8	17.1 ± 1.5
4	Mg-1.4BN	145 ± 3	217 ± 5	7.2 ± 0.8	115 ± 4	319 ± 4	17.6 ± 2.0
Disintegrated melt deposition method processed materials							
1	Pure Mg	120 ± 9	169 ± 11	6.4 ± 0.7	70 ± 8	234 ± 8	20.7 ± 0.9
2	Mg-0.3BN	133 ± 4	193 ± 7	8.6 ± 0.8	84 ± 9	275 ± 12	18.9 ± 0.7
3	Mg-0.6BN	154 ± 2	223 ± 2	15.3 ± 0.6	97 ± 3	297 ± 8	19.9 ± 1.2
4	Mg-1.2BN	178 ± 5	255 ± 3	12.6 ± 1.3	109 ± 4	307 ± 6	19.7 ± 1.4

Tensile Properties

Room temperature tensile properties of Mg/BN nanocomposites synthesized using the techniques of DMD and PM are provided in Table 4. The representative flow curves are shown in Fig. 5. The results indicate a noticeable improvement in mechanical properties in terms of strength and ductility due essentially to the addition of nanosized BN. With respect to processing method, while increasing the amount of nano-BN resulted in an improvement in ductility for the disintegrated melt deposition (DMD) processed materials, only a marginal increment in ductility was recorded for the powder metallurgy (PM) processed materials. This can be attributed to the higher porosity levels in the PM processed materials as shown in Table 2.

Compressive Properties

The room temperature compressive properties of PM and DMD processed Mg/BN nanocomposites are listed in Table 4. In case of the extruded Mg materials having a strong basal texture, the higher yield strength under tension and lower yield strength in compression (as seen in Table 4) is common due to differences in the deformation modes and initial crystallographic orientation. While a higher yield strength under tensile loading results from a difficulty in activation of basal slip coupled with an inability of twinning in the case of strong fiber texture. The ease of tensile twinning under compression loads results in lower compressive yield strength. Further, the presence of dispersed reinforcing phase in the Mg/BN nanocomposite would obstruct both twin nucleation and propagation, thereby restricting the twinning process to contribute towards higher strengths and resultant poor failure

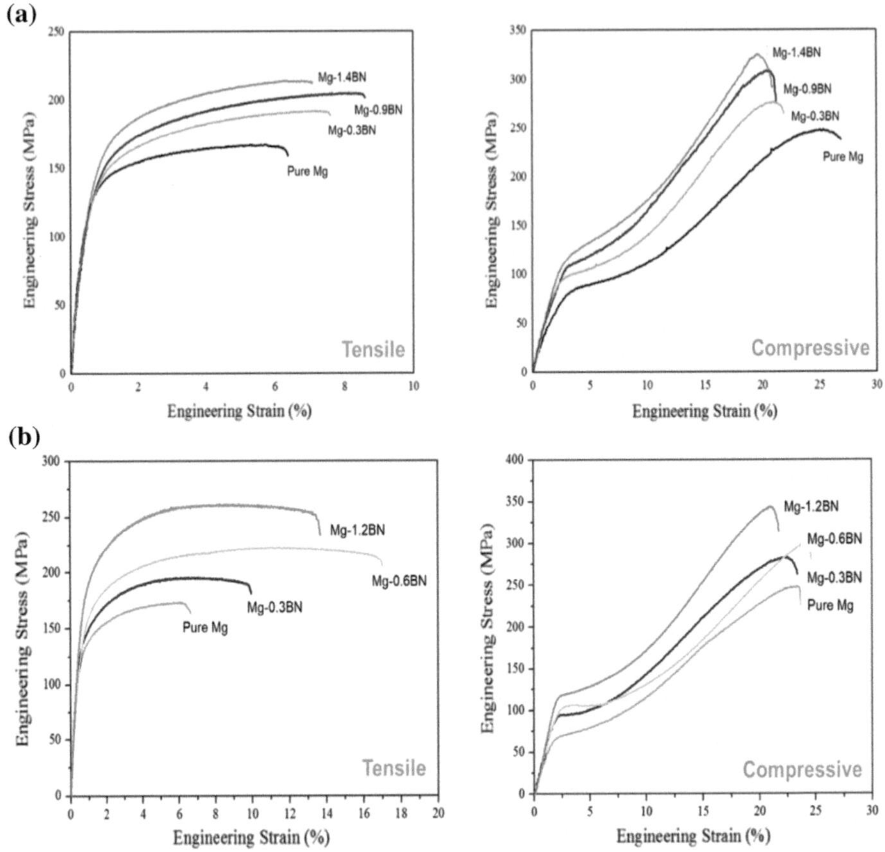

Fig. 5 Engineering stress versus strain curves of the powder metallurgy (PM) processed and disintegrated melt deposition (DMD) processed pure magnesium (Mg) and Mg/BN nanocomposites under tensile loading

strain. Between the powder metallurgy (PM) processed and disintegration melt deposition (DMD) processed Mg/BN nanocomposites, no significant difference in properties was observed since the key differentiating factor, i.e. porosity does not play a major role under compressive loading.

Conclusions

Magnesium (Mg) composites containing nanoscale BN particulates were successfully synthesized using both solid-state powder metallurgy and liquid-state disintegrated melt deposition methods. Based on a careful study essentially involving process-structure-property correlation, the following conclusions are drawn:

(a) The addition of nanoscale BN reinforcements to pure magnesium through powder metallurgy (PM) and disintegrated melt deposition (DMD) methods resulted in a marginal reduction in the average grain size and CTE.
(b) XRD studies conducted on Mg/BN nanocomposites that were developed using the techniques of PM and DMD showed strong basal texture, which is common for wrought magnesium (Mg) materials.
(c) Under tensile and compressive loads, both the PM and DMD processed Mg/BN nanocomposites revealed enhanced ultimate tensile strength and ductility. This is attributed to the fine distribution of nanoscale boron nitride (BN) particles through the microstructure coupled with associated contribution through dispersion strengthening.
(d) While the results of tensile testing indicated significant improvement in ductility for the DMD processed materials, only a marginal difference in properties was recorded under compressive loading.

References

1. Subramanian J, Seetharaman S, Gupta M (2015) Gupta processing and properties of aluminum and magnesium based composites containing amorphous reinforcement: a review. Metals (Basel) 5(2)
2. Tharumarajah A, Koltun P (2007) Is there an environmental advantage of using magnesium components for light-weighting cars? J Clean Prod 15(11–12):1007–1013
3. Mattes WG (2007) The BMW approach to tier 2 bin 5. In: 13th DEER conference, pp 1–29
4. Callister W, Rethwisch D (2007) Materials science and engineering: an introduction. 94
5. Kainer KU (2006) Basics of metal matrix composites
6. Jayalakshmi S, Gupta M (2015) Light metal matrix composites. In: Metallic amorphous alloy reinforcements
7. Gupta M, Sharon NML (2010) Magnesium, magnesium alloys, and magnesium composites
8. Nanotechnology for energy sustainability (2017)
9. Seetharaman S, Subramanian J, Tun KS, Hamouda AS, Gupta M (2013) Synthesis and characterization of nano boron nitride reinforced magnesium composites produced by the microwave sintering method. Materials (Basel) 6(5)
10. Sankaranarayanan S, Sabat RK, Jayalakshmi S, Suwas S, Almajid A, Gupta M (2015) Mg/BN nanocomposites: Nano-BN addition for enhanced room temperature tensile and compressive response. J Compos Mater 49(24)
11. Sankaranarayanan S, Habibi MK, Jayalakshmi S, Jia Ai K, Almajid A, Gupta M (2015) Nano-AlN particle reinforced Mg composites: microstructural and mechanical properties. Mater Sci Tech 31(9):1122–1131
12. Tun KS, Jayaramanavar P, Nguyen QB, Chan J, Kwok R, Gupta M (2012) Investigation into tensile and compressive responses of Mg–ZnO composites. Mater Sci Tech 28(5):582–588
13. Humphreys FJ, Hatherly M (2007) Grain growth following recrystallization. In: Recrystallization and related annealing phenomena
14. Tun KS, Gupta M (2008) Effect of extrusion ratio on microstructure and mechanical properties of microwave-sintered magnesium and Mg/Y2O3 nanocomposite. J Mater Sci 43(13):4503–4511

The Mechanical and Thermal Response of Shape Memory Alloy-Reinforced Aluminum Nanocomposites

M. Penchal Reddy, Vyasaraj Manakari, Gururaj Parande, R. A. Shakoor, T. S. Srivatsan and Manoj Gupta

Abstract The concept of developing intelligent materials that has the innate capability of healing its damage has engineered considerable scientific and even technological interest due on account of its potential for selection and use in sectors spanning aerospace, automotive, and even commercial products. Aluminium (Al) with its noteworthy properties, such as high specific strength (σ/ρ), low coefficient of thermal expansion, high thermal conductivity and good wear resistance characteristics is an ideal candidate for engineering the development of self-healing materials. In this paper, shape memory alloy [$Ni_{50}Ti_{50}$ (NiTi)]-reinforced aluminum matrix nanocomposites [referred to henceforth through the text as SMA-AMNCs] were fabricated using the technique of powder metallurgy followed by hybrid microwave sintering. The intrinsic influence of addition of nanosized particles of the chosen alloy (NiTi) on microstructural development, mechanical properties and even thermal properties of the chosen aluminum are examined. With the addition of nanoparticles of the shape memory alloy (NiTi), a noticeable improvement in hardness, ultimate compression/tensile strength [σ_{UTS}], yield strength [σ_{YS}], damping capacity (Q^{-1}) and damping loss rate (L) was observed, with a concurrent decrease in the values of failure strain (ε_f) and coefficient of thermal expansion (CTE). The observed increase in properties of the engineered nanocomposite as a consequence of contributions from intrinsic microstructural effects is neatly presented and briefly discussed.

M. Penchal Reddy (✉) · V. Manakari · G. Parande · M. Gupta (✉)
Department of Mechanical Engineering, National University of Singapore, 9 Engineering Drive 1, Singapore 117576, Singapore
e-mail: mpeprm@nus.edu.sg

M. Gupta
e-mail: mpegm@nus.edu.sg

R. A. Shakoor
Center for Advanced Materials, Qatar University, Doha 2713, Qatar

T. S. Srivatsan
Department of Mechanical Engineering, The University of Akron, Akron, OH 44325-3903, USA

T. S. Srivatsan and M. Gupta (eds.), *Nanocomposites VI: Nanoscience and Nanotechnology in Advanced Composites*, The Minerals, Metals & Materials Series, https://doi.org/10.1007/978-3-030-35790-0_4

Keywords Aluminum-shape memory alloy nanocomposites · Microwave sintering · Extrusion · Mechanical properties · Thermal property · Damping behavior

Introduction

A synergism of lightweight coupled with excellent specific mechanical properties of the aluminum (Al)-based materials has over a substantial period engineered significant attention and resultant action. With noticeable advances in technology coupled with urgent need for lightweight materials for the purpose of both selection and use in a spectrum of applications spanning the industries of automotive, aerospace and even commercial products, the addition of ceramic reinforcements to an aluminum matrix has been critically examined as a technically viable and economically affordable approach to replace the commercial alloys for use in a spectrum of performance-critical and even non-performance-critical applications. Ceramic particle-reinforced aluminum metal matrix composites (AMMCs) have been chosen for use in these sectors owing to their exceptional physical properties and mechanical properties, such as (i) lightweight, (ii) low cost, (iii) high specific modulus (E/ρ), (iv) high specific strength (σ/ρ) and (v) low coefficient of thermal expansion [1, 2]. Several ceramic materials have been chosen for use as the reinforcing phase for the aluminum-based composites, and a few of these include the following: (i) silicon carbide (SiC), (ii) aluminum oxide (Al_2O_3), (iii) boron nitride (BN), and (iv) titanium carbide (TiC) [3–5]. In an attempt to further enhance the mechanical properties, aluminum-based materials have been incorporated with nanoparticles with the prime objective of engineering a composite material. This has proven to simultaneously increase not only the strength but also the ductility of aluminum.

Current research studies have focused their attention on examining the influence of the addition of nanosized reinforcement particles on both thermal and mechanical behaviors of aluminum-based composites [6–8]. The engineered nanocomposites can be tailor-made to suit different applications by an appropriate selection of not only the reinforcement but also the fabrication technique used. In the recent past, metallic glass as the reinforcing phase was considerable importance from a technological perspective due on account of its high strength and hardness, enhanced resistance to corrosion coupled with acceptable to good functionality [9]. Various studies have convincingly shown that uniformly distributed metallic glass reinforcement nanoparticles can simultaneously improve both the strength and ductility of the aluminum-based composites [10, 11].

The $Ni_{50}Ti_{50}$ metallic glass alloy particles are a potential material for use as nanoparticles due to essence to their shape memory effect, high damping capacity coupled with super-elastic properties. A few of the earlier studies have attempted to present noticeable achievements in the mechanical properties of NiTi shape memory alloy-reinforced aluminum (Al) matrix, magnesium (Mg) matrix, and silver (Ag) matrix composites [12–16]. For the time being, NiTi shape memory alloy-reinforced aluminum matrix composites have been prepared using the techniques of (i) friction stir processing, (ii) ultrasonic additive manufacturing, (iii) pressure infiltration process, and (iv) spark plasma sintering [13–15, 17, 18]. Among the different synthesis methods used for the aluminum-based composites, the technique of powder metallurgy (PM) using an innovative, microwave sintering technique is particularly suitable for the synthesis of composite materials as it provides an excellent control over particle size growth, intrinsic microstructural changes coupled with careful control over volume fraction of the matrix and the reinforcement [19]. Microwave sintering of a NiTi alloy-reinforced aluminum matrix composite has not as yet been reported in the published literature.

To the best of our knowledge, no report has been published till date on aspects specific to a study and/or an investigation of the mechanical, thermal, and damping behaviors of Al–NiTi nanocomposite fabricated using the method of microwave sintering followed by hot extrusion. Therefore, the aim of the present research study was to investigate the microstructural evolution, mechanical, thermal, and damping behavior of aluminum nanocomposite reinforced with $Ni_{50}Ti_{50}$ shape memory alloy nanosized particles.

Experimental Procedure

Materials and Processing

Commercially available aluminum (Al) powder of size ~45 μm and purity of >99.7% [supplied by Merck (Darmstadt, Germany)] was used as the matrix material and the $Ni_{50}Ti_{50}$ shape memory alloy nanoparticles having an average size of ~30–50 nm with purity >99 + % [procured from nanostructured and amorphous materials, Inc. Houston, TX, USA] were used as the reinforcement phase to develop the Al–NiTi nanocomposites.

Details specific to the preparation process of the Al–NiTi nanocomposites have been reported in our earlier study [8]. To produce the Al–NiTi nanocomposites, 1.5 volume pct. of NiTi alloy powder and pure aluminum were precision weighed in an electronic balance and subsequently blended at 200 rpm for 2 h (120 min) using a planetary ball mill [*Model*: Retsch PM400]. The blended composite powder was then compacted into cylindrical billets that had a height of 40 mm and a diameter of 35 mm at a pressure of 97 bars (50 tons). The compacted billets were then sintered using microwave sintering at 550 °C under ambient conditions. Billets of the microwave sintered aluminum and aluminum nanocomposite were homogenized at 400 °C for 1 h (60 min) prior to hot extrusion at a temperature of 350 °C at an extrusion ratio of 20.25:1 on a 150-ton hydraulic press resulting in rods that measured 8 mm in diameter. Samples of the as-synthesized composite samples were prepared for further characterized in accordance with procedures detailed in the ASTM standards.

Materials Characterization

X-ray diffraction (XRD) analysis was carried out on polished surfaces of the extruded aluminum sample and the aluminum nanocomposite sample using an automated diffractometer [*Model*: Shimadzu LAB-X XRD-6000]. Microstructural characterization of the extruded pure Al and Al–NiTi nanocomposite was made possible using a scanning electron microscope [*Model*: FESEM-S4300, HITACHI LTD., Japan].

Microhardness measurements were performed on polished surfaces of the test samples using an automatic digital Vickers microhardness (HV) tester. Hardness of the test samples was determined using a load of 25 gf for a dwell time of 15 s. Room temperature uniaxial compressive tests and tensile tests were performed on both the pure aluminum and composite samples in accordance with procedures detailed in the standards ASTM E9-89a and ASTM E8/E8 M-15a, using a Lloyd universal testing machine (LR 50KN). The samples were deformed at a strain rate of 8.3×10^{-4}/sec for all tests.

The damping performance of pure Al and Al–NiTi nanocomposite was conducted according to the ASTM standard E1876-09 and using a resonance frequency damping analyzer [Model: RFDA, IMEC, Genk, Belgium]. Each sample used for this specific study measured 60 mm in length and 8 mm in diameter.

The coefficient of thermal expansion (CTE) of the extruded aluminum and the engineered composites was determined using a thermomechanical analyzer [*Model*: SETARAM 92-16/18]. A heating rate of 5 °C/min was maintained with a flow rate of argon gas at 0.1 L per minutes (lpm). With the aid of an alumina probe, the displacement experienced by the test samples was precision measured as a function of temperature.

Results and Discussion

Microstructure

Visual observation of the microwave sintered nanocomposite billet and the extruded rods clearly revealed an absence of any macroscopic surface defects. This confirms an overall suitability of the processing parameters used during the processes of sintering and extrusion. The intrinsic microstructural features of the developed pure aluminum and Al–1.5 NiTi nanocomposite were studied in terms of the presence and distribution of the reinforcing NiTi nanoparticles through the microstructure. Figure 1 shows the results of the Al–1.5 NiTi nanocomposite. It is observed that the NiTi particles are homogeneously distributed through the aluminum matrix. Further, the extrusion process provided for a suitable adhesion between the aluminum particles while concurrently enabling the shape memory alloy particles to be distributed homogeneously through the aluminum metal matrix, which made the secondary process much more effective in engineering a near-uniform composite.

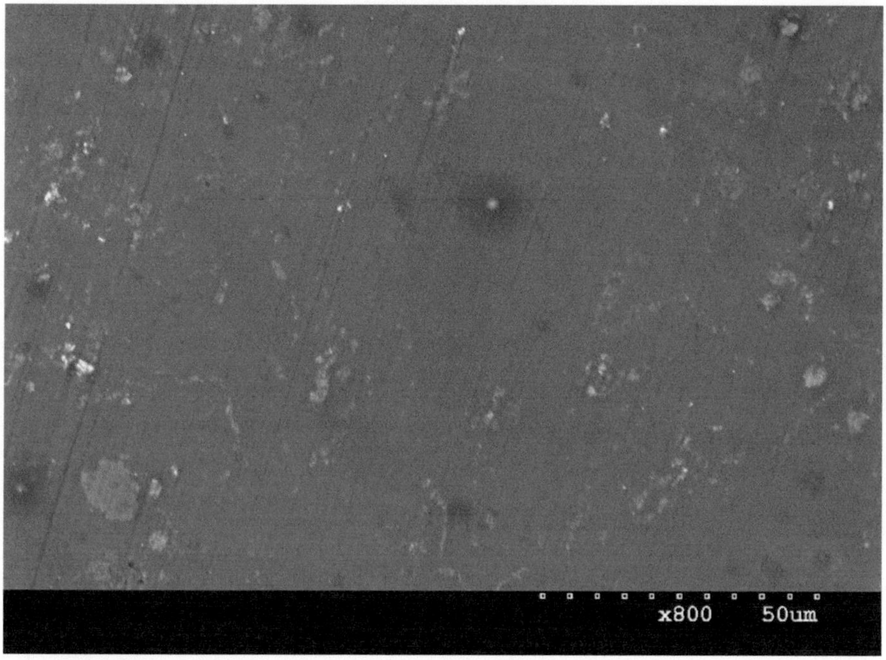

Fig. 1 Distribution of NiTi nanoparticles in the Al–1.5 NiTi nanocomposite

Phase Analysis

X-ray diffraction is a material characterization technique that can be useful for the purpose of analyzing any textural changes to the composite subsequent to secondary processing. The X-ray diffraction (XRD) patterns of extruded pure Al and the Al–1.5 NiTi nanocomposite are shown in Fig. 2. The overall appearance of the reinforcing NiTi particles in the aluminum matrix is made possible. The X-ray diffraction patterns exhibit the peaks of aluminum to occur at 38° (111), 44° (200), 65° (220), 78° (311) and 83° (222) and a few weak peaks of the reinforcing NiTi phase at 42° (110) and 79° (211). These are the characteristic peaks and in agreement with the peaks observed and reported by Czeppe and co-workers [20].

Microhardness Measurements

Results of hardness measurements of pure Al and Al–1.5 NiTi nanocomposite are summarized in Table 1. It is observed that the Al–1.5 NiTi nanocomposite possesses a high hardness value of 78 $\pm$ 5 HV, whereas the average hardness of pure Al is about 38 $\pm$ 4 Hv. The observed improvement in the hardness value can be attributed to the

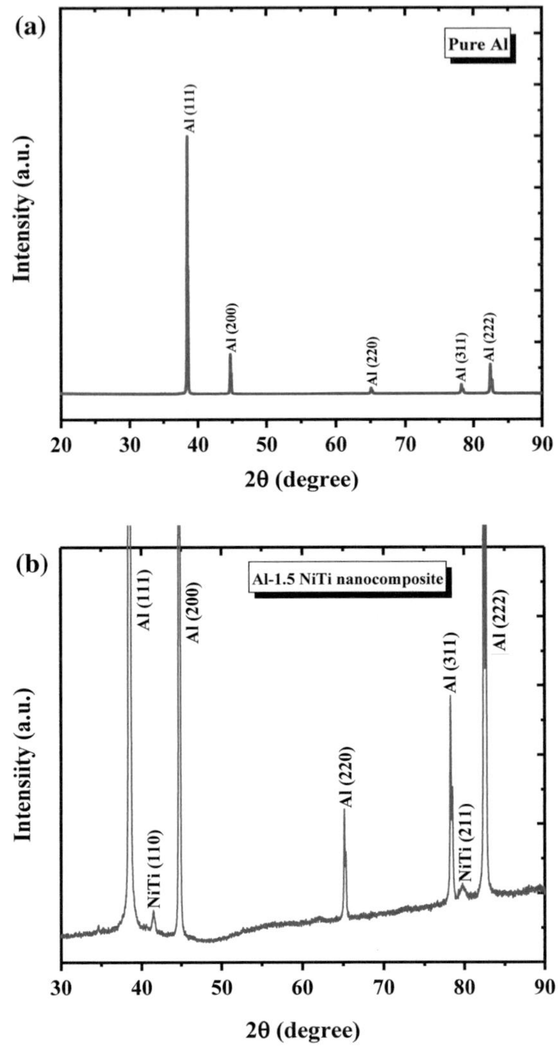

Fig. 2 X-ray diffraction pattern of **a** extruded pure Aluminum, and **b** extruded Al–1.5 NiTi nanocomposite

Table 1 Results of microhardness measurements

Sample	Hardness (Hv)
Pure Al	38 ± 4
Al–1.5 vol.% NiTi	78 ± 5

conjoint and mutually interactive influences of the following phenomenon occurring at the microscopic level [4]:

Table 2 Compressive properties of pure Al and Al–NiTi nanocomposite

Sample	Compressive properties		
	CYS (MPa)	UCS (MPa)	Failure strain (%)
Pure Al	70 ± 3	263 ± 4	9.5 ± 0.2
Al–1.5 vol.% NiTi	97 ± 4	385 ± 5	6.4 ± 0.4

(a) An incorporation of high hardness NiTi nanoparticles,
(b) Dispersion hardening effect provided by the nanoparticles in the soft and ductile metal matrix, and
(c) Strengthening effect resulting from constricting the growth of the grains in pure aluminum.

Compressive Behavior

Results of the compression test are summarized in Table 2. The compressive yield strength (CYS) of pure aluminum increased from 70 ± 3 to 97 ± 4 MPa with the addition of 1.5 vol. % NiTi nanoparticles.

The Al–1.5 NiTi nanocomposite exhibited an ultimate compressive strength (UCS) of 385 ± 3 MPa, which is ~46% more than that of pure aluminum (263 ± 4). However, fracture strain of the nanocomposite sample decreased to 6.4% when compared one-on-one with pure aluminum, which exhibited a fracture strain of 9.5%. The observed increase in the 0.2% compressive yield strength (CYS) and ultimate compressive strength (UCS) of the Al–NiTi nanocomposite can be attributed to the conjoint and mutually interactive influences of the following mechanisms occurring at the fine microscopic level:

(a) Presence of fairly dispersed NiTi particles [21],
(b) Effective transferring of load from the Al matrix to the reinforcing NiTi nanoparticles [22],
(c) Mismatch in elastic modulus and coefficient of thermal expansion values of the soft and ductile metal matrix and the hard and brittle reinforcement resulting in the generation of dislocations [23], and
(d) An Orowan strengthening effect due to the presence of hard, brittle and essentially elastically deforming reinforcing NiTi nanoparticles [23].

Tensile Behavior

The room temperature tensile properties of the Al and Al–NiTi nanocomposite are provided in Table 3, and the stress–strain curves are shown in Fig. 3. An incorporation

Table 3 Results of room temperature tensile properties

Sample	Tensile properties		
	TYS (MPa)	UTS (MPa)	Elongation (%)
Pure Al	106 ± 3	119 ± 3	10.8 ± 0.5
Al–1.5 vol.% NiTi	154 ± 3	168 ± 5	5.5 ± 0.2

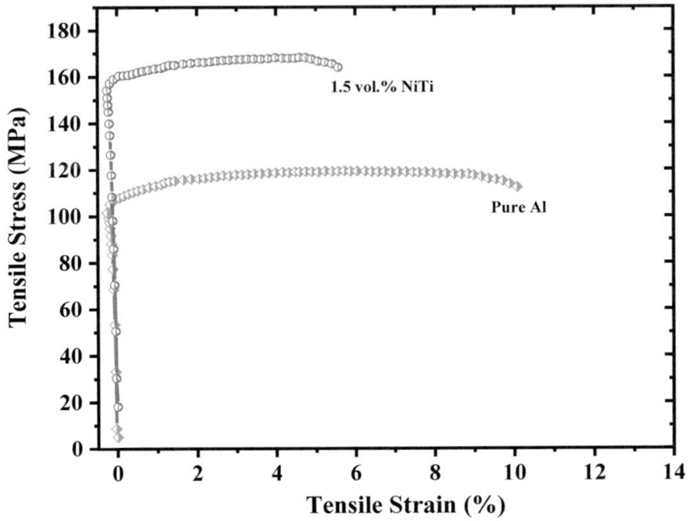

Fig. 3 Room temperature tensile stress–strain curves of extruded pure Al and the engineered composite

of the reinforcing NiTi nanoparticles into the pure Al metal matrix led to an improvement in both the 0.2% tensile yield strength (TYS) and ultimate tensile strength (UTS), whereas the tensile elongation decreased. The Al–1.5 NiTi nanocomposite exhibited the maximum tensile yield strength (TYS) and ultimate tensile strength (UTS) values of 154 ± 3 MPa and 168 ± 5 MPa, which is ~ 45% and 41% greater than that of pure aluminum. The key factors contributing to the observed improvement in strength could be due to the load transfer mechanism coupled with an increase in dislocation density due to noticeable differences in the values of CTE and elastic modulus between the pure, soft and ductile Al and the hard and brittle reinforcing NiTi nanoparticles. As for the tensile failure strain (ε_f), a decrease in failure strain with an incorporation of the NiTi nanoparticles was observed. These results are quite consistent with the results of studies made on the following: (i) nanocomposites synthesized using the powder metallurgy technique followed by conventional hot extrusion, and (ii) spark plasma sintered aluminum matrix composites [17, 18].

Table 4 Results of coefficient of thermal expansion

Sample	CTE ($\times 10^{-6}$/K)
Pure Al	23.65
Al–1.5 vol.% NiTi	22.8

Coefficient of Thermal Expansion

Results of coefficient of thermal expansion (CTE) measurements made on pure aluminium and nanocomposite sample is provided in Table 4. The CTE value for the Al–1.5 NiTi nanocomposite is (22.8×10^{-6}/K) lower than that of pure aluminum (23.65×10^{-6}/K). The test results reveal that the observed reduction in CTE of aluminium is a result of the addition and presence of nanosized particles of the shape memory alloy.

Damping Behavior

The vibration damping capabilities of both pure Al and the Al–1.5 NiTi nanocomposite are shown in Fig. 4 and summarized in Table 5. An enhancement in the damping capacity (Q^{-1}) and damping loss rate (L) of the pure aluminum Al is observed with the addition of NiTi nanoparticles. The Al–1.5 NiTi nanocomposite revealed the best

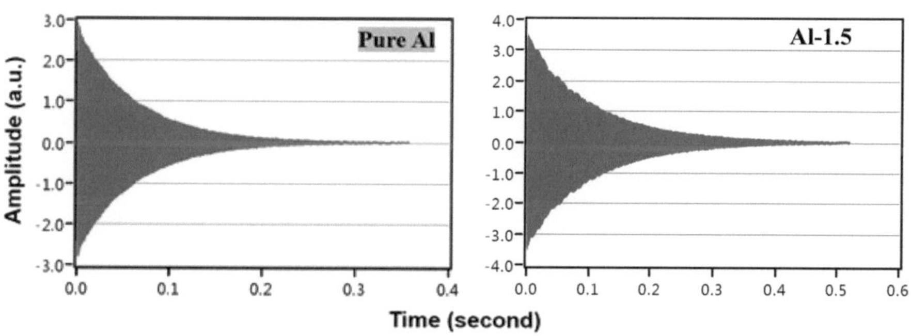

Fig. 4 Damping characteristics of pure Al and Al–1.5 vol.% NiTi nanocomposite

Table 5 Elastic modulus and damping characteristics of the extruded Al–NiTi nanocomposites

Composition	Damping capacity ($\times 10^{-4}$)	Damping loss rate	Elastic modulus (GPa)
Pure Al	5.33 ± 0.056	14.76 ± 0.09	71.65 ± 0.02
Al–1.5 vol.% NiTi	6.15 ± 0.029 (↑15.4%)	17.15 ± 0.06 (↑16.2%)	75.93 ± 0.01 (↑6%)

values of damping capacity and damping loss rate to be ~6.15 and ~17.15 × 10^{-4}, respectively. The damping capacity and damping loss rate for the nanocomposite sample improved by ~15.4% and ~16.2% when compared one-on-one with that of pure aluminum (Al).

Several damping-related mechanisms are responsible for the observed improvement in the damping characteristics of pure aluminum (Al) with the addition of NiTi nanoparticles and these include the following:

(a) Bulk texture modification,
(b) Enhanced damping at the interface binding the NiTi nanoparticles in the soft aluminum (Al) matrix, and
(c) High number of dislocations owing to mismatch in the thermal expansion coefficient of the soft and ductile aluminum (Al) matrix and the hard and brittle reinforcing NiTi nanoparticles.

The damping response of aluminum-based materials can also be influenced by the presence of microscopic defects and porosity. These mechanisms are primarily the result of a combined interaction between these phenomena and not a characteristic effect of each. Further, from Table 5, it is observed that the elastic modulus [E] of the nanocomposite sample increased with the addition of NiTi nanoparticles. This is consistent with the trend exhibited by aluminum (Al)-based metal matrix composites [8] indicating a good interface between the reinforcing NiTi nanoparticles and the soft and ductile Al matrix.

Conclusions

In this research study, pure aluminum (Al) and Al–1.5 vol.% NiTi nanocomposite were fabricated using microwave sintering followed by hot extrusion process. The microstructure, mechanical, thermal and damping performance were investigated. The key finds are the following:

(a) Scanning electron microscopy (SEM) and X-ray diffraction (XRD) results indicate that nano-NiTi alloy particles to be homogeneously distributed through the Al metal matrix.
(b) The maximum hardness value obtained was 110% higher than that of pure Al sample.
(c) Compression tests revealed an increase in the strength from 263 ± 4 MPa for pure Al to 385 ± 5 MPa (~46%) for the Al–1.5 NiTi nanocomposite.
(d) Tensile strength of the samples was improved from 119 ± 3 MPa for pure Al to 168 ± 5 MPa (~41%) for the Al–1.5 NiTi nanocomposite.
(e) The observed increase or enhancement in the mechanical properties can be attributed to the conjoint influence of secondary processing, homogenously distributed particles and dispersion strengthening effect.

(f) Coefficient of thermal expansion showed a reverse trend indicating an increase in thermal stability.
(g) The Al–1.5 NiTi nanocomposite showed the best damping (damping loss rate, damping capacity and elastic modulus) response.

The technique of microwave sintering has potentially wider range of advantages and applications in the preparation of aluminum (Al)-based nanocomposites tools primarily because of the advantages associated with the heating mechanism.

References

1. Matli PR, Ubaid F, Shakoor RA, Parande G, Manakari V, Yusuf M, Mohamed AMA, Gupta M (2017) Improved properties of Al–Si_3N_4 nanocomposites fabricated through a microwave sintering and hot extrusion process. RSC Adv 7(55):34401–34410
2. Zhang Y, Ma N, Wang H, Le Y, Li X (2007) Damping capacity of in situ TiB_2 particulates reinforced aluminium composites with Ti addition. Mater Des 28(2):628–632
3. Jayalakshmi S, Gupta S, Sankaranarayanan S, Sahu S, Gupta M (2013) Structural and mechanical properties of $Ni_{60}Nb_{40}$ amorphous alloy particle reinforced Al-based composites produced by microwave-assisted rapid sintering. Mater Sci Eng, A 581:119–127
4. Reddy MP, Shakoor R, Parande G, Manakari V, Ubaid F, Mohamed A, Gupta M (2017) Enhanced performance of nano-sized SiC reinforced Al metal matrix nanocomposites synthesized through microwave sintering and hot extrusion techniques. Progress Natural Sci: Mater Int 27(5):606–614
5. Reddy MP, Ubaid F, Shakoor R, Parande G, Manakari V, Mohamed A, Gupta M (2017) Effect of reinforcement concentration on the properties of hot extruded Al-Al_2O_3 composites synthesized through microwave sintering process. Mater Sci Eng, A 696:60–69
6. Moazami-Goudarzi M, Akhlaghi F (2016) Wear behavior of Al 5252 alloy reinforced with micrometric and nanometric SiC particles. Tribol Int 102:28–37
7. Reddy MP, Himyan M, Ubaid F, Shakoor R, Vyasaraj M, Gururaj P, Yusuf M, Mohamed A, Gupta M (2018) Enhancing thermal and mechanical response of aluminum using nanolength scale TiC ceramic reinforcement. Ceram Int 44(8):9247–9254
8. Reddy MP, Manakari V, Parande G, Ubaid F, Shakoor R, Mohamed A, Gupta M (2018) Enhancing compressive, tensile, thermal and damping response of pure Al using BN nanoparticles. J Alloy Compd 762:398–408
9. Byrne CJ, Eldrup M (2008) Bulk metallic glasses. Science 321(5888):502–503
10. Reddy MP, Ubaid F, Shakoor R, Mohamed A (2018) Microstructure and mechanical behavior of microwave Sintered Cu 50 Ti 50 amorphous alloy reinforced Al metal matrix composites. JOM 70(6):817–822
11. Scudino S, Surreddi K, Sager S, Sakaliyska M, Kim J, Löser W, Eckert J (2008) Production and mechanical properties of metallic glass-reinforced Al-based metal matrix composites. J Mater Sci 43(13):4518–4526
12. Chau E, Friend C, Allen D, Hora J, Webster J (2006) A technical and economic appraisal of shape memory alloys for aerospace applications. Mater Sci Eng, A 438:589–592
13. Hao S, Cui L, Jiang J, Guo F, Xiao X, Jiang D, Yu C, Chen Z, Zhou H, Wang Y (2014) A novel multifunctional NiTi/Ag hierarchical composite. Sci Reports 4:5267
14. Ni D, Wang J, Zhou Z, Ma Z (2014) Fabrication and mechanical properties of bulk NiTip/Al composites prepared by friction stir processing. J Alloy Compd 586:368–374
15. Parande G, Manakari V, Wakeel S, Kujur M, Gupta M (2018) Enhancing mechanical response of Monolithic Magnesium Using Nano-NiTi (Nitinol) particles. Metals 8(12):1014

16. Sankaranarayanan S, Shankar VH, Jayalakshmi S, Bau NQ, Gupta M (2015) Development of high performance magnesium composites using $Ni_{50}Ti_{50}$ metallic glass reinforcement and microwave sintering approach. J Alloy Compd 627:192–199
17. Hu J, Zhang Q, Wu G, Liu Y, Li D (2014) Effect of pre-oxidation of TiNi fibers on the interfacial and mechanical property of TiNif/Al composite. Mater Sci Eng, A 597:20–28
18. Wang Z, Dong P, Wang W, Yan Z, Ding M (2017) Interface formation of TiNif/Al composites fabricated by spark plasma sintering. J Alloy Compd 726:507–513
19. Thakur SK, Tun KS, Gupta M (2010) Enhancing uniform, nonuniform, and total failure strain of aluminum by using SiC at nanolength scale. J Eng Mater Technol 132(4):041002
20. Czeppe T, Levintant-Zayonts N, Swiatek Z, Michalec M, Bonchyk O, Savitskij G (2009) Inhomogeneous structure of near-surface layers in the ion-implanted NiTi alloy. Vacuum 83:S214–S219
21. Parande G, Manakari V, Meenashisundaram GK, Gupta M (2016) Enhancing the hardness/compression/damping response of magnesium by reinforcing with biocompatible silica nanoparticulates. Int J Mater Res 107(12):1091–1099
22. Nguyen Q, Gupta M (2008) Enhancing compressive response of AZ31B magnesium alloy using alumina nanoparticulates. Composites Science and Technology 68(10–11):2185–2192
23. Hassan S, Gupta M (2003) Development of high strength magnesium copper based hybrid composites with enhanced tensile properties. Mater Sci Technol 19(2):253–259

Electrical Performance of Bulk Al–ZrB_2 Nanocomposites from 2 K to 300 K

Shuaihang Pan, Gongcheng Yao, Jie Yuan and Xiaochun Li

Abstract Electrical properties are of significance for metals/alloys and their applications. While nanoparticles can enhance mechanical performance of metals/alloys, there is a strong need to understand how nanoparticles affect their electric behavior at various temperatures. In this study, ZrB_2 nanoparticles were synthesized in situ to cast bulk Al–ZrB_2 samples for electric characterizations. The electrical conductivity, electron mobility, and electron concentration of Al–3 vol.% ZrB_2 were measured in the temperature range from 2 K to 300 K. The effects of in situ ZrB_2 nanoparticles on the Al matrix were systematically studied in terms of its compositions, morphologies, grain sizes, and nanophase sizes. It is discovered the Al–ZrB_2 interfaces play a key role in tuning structural and electrical performances. This mechanism is important to better understand the electron behaviors in Al alloys containing in situ nanoparticles. The in situ fabrication and electrical characterization methods can be readily applied to other metallic nanocomposites.

Keywords Electrical conductivity · Aluminum · ZrB_2 · Nanocomposites · Low-temperature measurement

Introduction

Metals and alloys have unique electrical properties due to a high concentration of free electrons. In various electronic products [1, 2], thermal management applications [3], and conducting fields [4], the electrons dominate the performance of metals and alloys [5]. Thus, understanding the electron behavior is critical for a rational design of functional metallic systems.

S. Pan · X. Li (✉)
SciFacturing Laboratory, School of Mechanical and Aerospace Engineering, University of California, Los Angeles, CA 90095, USA
e-mail: xcli@seas.ucla.edu

G. Yao · J. Yuan · X. Li
School of Materials Science and Engineering, University of California, Los Angeles, CA 90095, USA

T. S. Srivatsan and M. Gupta (eds.), *Nanocomposites VI: Nanoscience and Nanotechnology in Advanced Composites*, The Minerals, Metals & Materials Series, https://doi.org/10.1007/978-3-030-35790-0_5

Recently, metal matrix nanocomposites (MMNCs) are emerging as a novel class of materials to offer metals and alloys supreme mechanical properties [6]. While the strengthening [6, 7] and phase modification effects [8] from nanoparticles like ZrB_2, TiB_2, and TiC are well studied, it remains unknown how the electrons affect the electrical performance of these novel MMNCs. In this case, while expecting a good combination of mechanical and electrical properties in MMNCs [9], it is important to resolve the dilemma with systematic analyses into the electron behavior.

To achieve this goal, many studies of metal matrix nanocomposites consider Al as a model system, because its alloys are widely used as electrical structural materials [10–12] and wire conductors. It's noticeable that many efforts have been devoted to clarify the effects of nanophase percentage [7, 13] and size [14]. Thereafter, useful theories including effective medium theory (EMT) have been developed to describe the electron behavior [13, 14]. The main problem of these investigations resides in the following several aspects: First, though nanophase percentage and size are important for the depiction of collective electron behavior (e.g. room temperature electrical conductivity), the true interaction between Al matrix with the nanophase is not clear. Second, many experiments interpret the electrical performance from configurational factors including dislocations, grain boundaries, etc. [9, 14, 15]; the electronic parameters like electron mobility are not quantitatively discussed; and the correlation between configurational and electronic factors is missing. Last but not the least, experiments directly revealing the relationships between these electronic parameters have not been conducted and the temperature-dependence electrical performance has not been studied in Al–MMNCs. Therefore, a wide temperature scanning for Al–MMNC electrical properties and an investigation of inter-connected electronic parameters are urgently needed.

In this paper, in situ synthesized 3 vol.% ZrB_2 are used as the nanophase in Al matrix. Microstructures including grain size and nanoparticle size are discussed to separate their effects on electrical conductivity. By measuring the electrical conductivity from 2 K to 300 K, it is confirmed that the metallic characteristics of electrical properties remained in the Al–3 vol.% ZrB_2 nanocomposites. By measuring the electron mobility and (apparent) free electron concentration, the electronic contribution to the reduction of electrical conductivity in Al–3 vol.% ZrB_2 is quantitatively analyzed. With configurational factors, a complete picture of how ZrB2 nanoparticles affect the matrix electrical performance is obtained. This study is of great significance to the design of Al–MMNCs for desired electronic applications.

Materials and Methods

The Al–3 vol.% ZrB2 nanocomposites were prepared via an in situ molten salt assisted method following the following reaction route:

$3K_2ZrF_6 + 6KBF_4 + 10Al \rightarrow 2KF + 10KAlF_4 + 3ZrB_2$ (750°C, 20 min)

3 vol.% ZrB_2 is selected for good castability of samples. The ZrB_2 nanoparticle size and distribution in Al were studied via scanning electron microscopy (SEM),

and the result is shown in Fig. 1. It's clear that the ZrB_2 nanoparticles have a pseudo-dispersion (Fig. 1 (a)) but without agglomeration or sintering (Fig. 1b) [6, 7]. The diameter of the synthesized ZrB_2 nanoparticles is ~112 nm on average. For the grain size, the samples were etched by a mixture of 50 mL Poulton's reagent, 25 mL HNO_3 (aq), and 40 mL chromic acid (3 g/10 mL distilled water) as the dye for 5–10 s [16]. The etched surface with exposed grain sizes was analyzed under polarized light. As shown in Fig. 2, with the incorporation of ZrB_2 nanoparticles, the grain size of the Al matrix decreases from ~1100 um in pure Al to about 179 um in the Al–3 vol.% ZrB_2 nanocomposite (Fig. 2c).

The sample for electrical conductivity measurement was machined by wire EDM into Hall bar shapes with a thickness of ~100–150 um, and then measured on physical property measurement system (PPMS) to reveal the low-temperature electrical properties with a 50 mA current pump under 8 Hz. The temperature scanning starts from 2 K up to 300 K, covering the reported possible ZrB_2 superconductivity transition temperature [17]. The magnetic scanning in PPMS for the apparent carrier (i.e. free

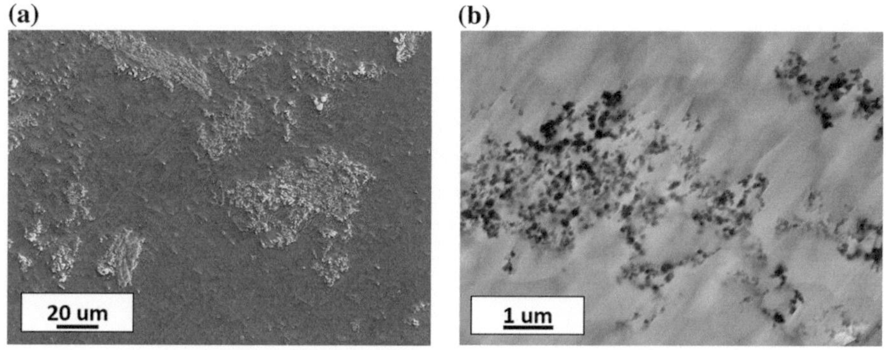

Fig. 1 **a** The microstructure for Al–3 vol.% ZrB_2 nanocomposites; **b** The ZrB_2 nanoparticle size distribution in the Al–3 vol.% ZrB_2 nanocomposites

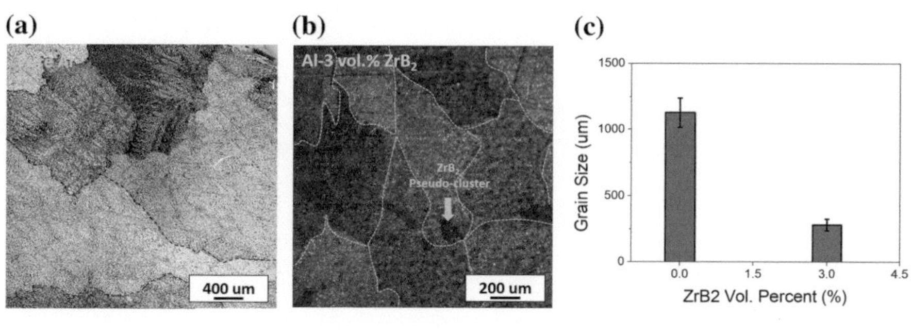

Fig. 2 The grain structure after etching for **a** pure Al and **b** Al–3 vol.% ZrB_2 nanocomposites; **c** The grain size comparison for pure Al and Al–3 vol.% ZrB_2 nanocomposites

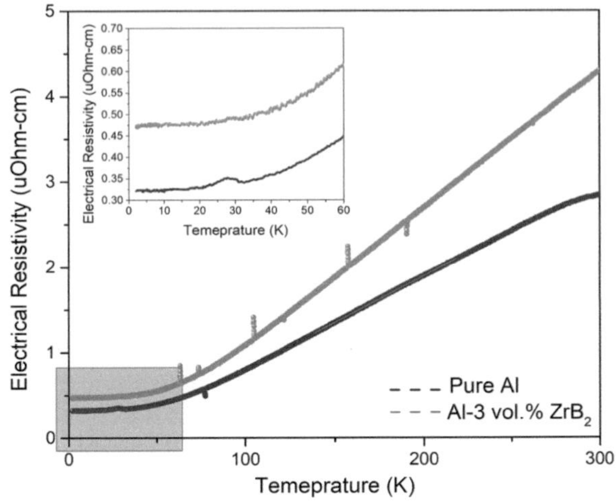

Fig. 3 Electrical conductivity in pure Al and Al–3 vol.% ZrB_2 nanocomposite from 2–300 K. (Inserted: The detailed scanning for electrical conductivity from 2 to 50 K)

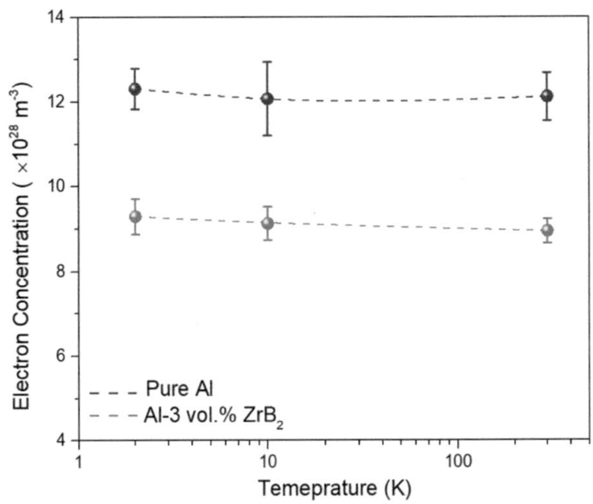

Fig. 4 Electron concentration of pure Al and Al–3 vol.% ZrB_2 at 2, 10, and 300 K

electron) density and electron mobility is in the range of ± 6.0 T [18]. The results are summarized in Figs. 3, 4 and 5 and Table 1, respectively.

Results and Discussion

The electrical conductivity of pure Al and Al–3 vol.% ZrB2 nanocomposites from 2 to 300 K is summarized in Fig. 3. At room temperature, the electrical resistivity of Al–3 vol.% ZrB2 is 4.1 μΩ·cm, larger than 2.6 μΩ·cm in pure Al. However, Al–3

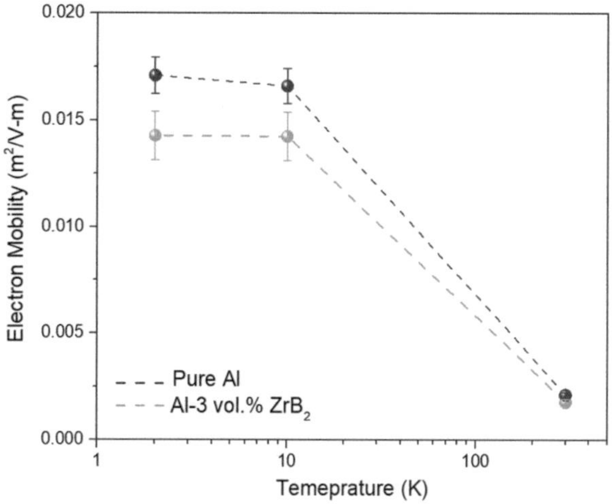

Fig. 5 The electron mobility of pure Al and Al–3 vol.% ZrB_2 at 2, 10, and 300 K

Table 1 Reported superconductivity temperature and measured Bloch–Grueneisen factor for Al and Al–3 vol.% ZrB_2[17]

Materials	Reported superconductivity temp. at 0T K	Bloch–Grueneisen factor μΩ·cm
Pure Al	1.2	4.26
Al–3 vol.% ZrB2	5.6 for ZrB_2	6.33

vol.% ZrB2 still shows characteristic metallic electron behavior following Eq. 1, as the electrical conductivity increases with the decreased temperature and saturates when the temperature approaches 0 K.

$$\Delta\rho(T) \propto T^5 \cdot \int_0^{\frac{T_\theta}{T}} \frac{x^5}{(e^x - 1)\cdot(1 - e^{-x})} \mathrm{d}x \tag{1}$$

Moreover, when gauging the electron concentration at 2, 10, and 300 K, it's found that the electron concentration has little change in both pure Al (~$12 \times 10^{28}\mathrm{m}^{-3}$) and Al–3 vol.% ZrB_2 nanocomposites (~$9 \times 10^{28} m^{-3}$). This confirms that the metallic collective behavior of electrons in Al–3 vol.% ZrB_2 is kept.

However, when the temperature is lower, the electron mobility is higher in both pure A and the Al–3vol.%ZrB_2 materials. Furthermore, for pure Al and Al–3 vol.% ZrB_2 at 2, 10, and 300 K, the electron mobilities are comparable, especially when the temperature is high and when the phonon is not suppressed. Moreover, at 2 K, the electron mobility does not suffer a steep drop due to the possible electron localization in the Al–3 vol.% ZrB_2 nanocomposites, implying that the reported superconductivity in ZrB2 plays little role in the nanocomposites.

When we calculated the Bloch–Grueneisen factor in terms of the temperature dependence of their electrical conductivity (See Table 1), the Al–3 vol.%

ZrB2 nanocomposites have a higher value than pure Al, which indicates a larger electron–phonon interaction after ZrB_2 nanoparticles are introduced.

To separate the effects from configurational and electronic contributions, the electrical resistivity from grain boundary and dislocation change is calculated for Al–3 vol.% ZrB2 nanocomposites via the following Eqs. 2 and 3: [19]

$$\Delta\rho_{GB} \cong \frac{2}{3} \cdot \alpha_T \cdot \Omega_{Al-GB} \cdot \left(\frac{2.37}{D}\right) \quad (2)$$

$$\Delta\rho_{dis} = 6 \cdot \Omega_{Al-dis} \cdot \frac{f \cdot \Delta CTE \cdot \Delta T}{r \cdot b \cdot (1-f)} \quad (3)$$

where Ω_{Al-GB} and Ω_{Al-dis} refer to the grain boundary resistance (1.9×10^{-16} $\Omega \cdot m^2$) [20] and dislocation resistance (3.3×10^{-25} $\Omega \cdot m^2$) [21]. f, D, r, and b are the volume percentage of ZrB_2, the grain size of Al, the radius of a ZrB_2 nanoparticle, and the Burgers vector for Al (0.2862 nm), respectively. ΔT refers to the temperature difference between the in situ synthesis temperature and the compared temperature.

By the comparison shown in Fig. 6, it is clear that, with a low volume percentage of ZrB_2 nanoparticles in Al, the configurational contributions by changing grain boundaries and dislocations are not directly correlated with the reduction of electrical conductivity in the Al–3 vol.% ZrB_2 nanocomposites. It could infer that the electrical conductivity reduction is larger than that of the electron concentration. It indicates that the electrical performance is affected via the indirect effect of the configurational defects on the electron concentration and electron mobility.

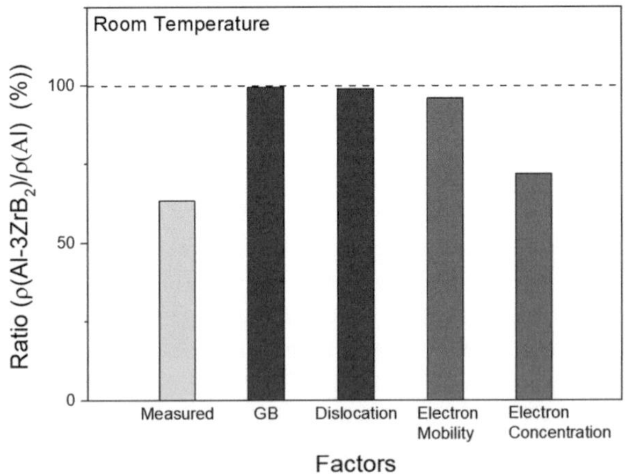

Fig. 6 Comparison between the factors contributing to the reduction of electrical conductivity in Al–3 vol.% ZrB_2 nanocomposites

Conclusions

This study provides a quantitative study into the electrical behavior of Al–ZrB_2 nanocomposites. The 3 vol.% ZrB_2 nanoparticles are synthesized in situ in the molten Al. The nanoparticles are pseudo-dispersed, and the Al grains are refined. After obtaining the electrical conductivity, electron mobility, and electron concentration via Hall measurement, the contributions to the reduced electrical conductivity from configurational (i.e. grain boundary, and dislocation) and electronic (i.e. electron mobility, and electron concentration) aspects are separated and quantitatively analyzed. It is confirmed that the altered grain boundaries and dislocations are not the major reason for the reduction of electrical conductivity in the Al–3 vol.% ZrB_2 nanocomposites. The major role of the incorporated ZrB_2 nanoparticles is to introduce the abovementioned defects that increase the phonon–electron interactions, thus leading to a significant decrease in (apparent) free electron concentration [22]. This understanding can help eliminate the errors and inaccuracy from determining MMNC electrical conductivity purely by configurational considerations. Moreover, a quantitative determination of electronic parameters in Al–ZrB_2 can guide the rational design in MMNC materials for electrical applications.

Acknowledgements This work was supported by the National Science Foundation and MetaLi L.L.C.

References

1. Pan S, Zhang Z (2018) Fundamental theories and basic principles of triboelectric effect: a review. Friction 1–16. https://doi.org/10.1007/s40544-018-0217-7
2. Guan Z, Hwang I, Pan S, Li X (2018) Scalable manufacturing of AgCu^40 (Wt.%)–WC nanocomposite microwires. J Micro Nano-Manuf. https://doi.org/10.1115/1.4040558
3. Pan S, Yao G, Sokoluk M, Guan Z, Li X (2019) Enhanced thermal stability in Cu-40 wt% Zn/WC nanocomposite. Mater Des 107964. https://doi.org/10.1016/j.matdes.2019.107964
4. Holwech I, Jeppesen J (1967) Temperature dependence of the electrical resistivity of Aluminium films. Philos Mag J Theor. Exp Appl Phys 15(134):217–228. https://doi.org/10.1080/14786436708227694
5. Gall D (2016) Electron mean free path in elemental metals. J Appl Phys 119(8):085101. https://doi.org/10.1063/1.4942216
6. Chen L-Y, Xu J-Q, Choi H, Pozuelo M, Ma X, Bhowmick S, Yang J-M, Mathaudhu S, Li X-C (2015) Processing and properties of magnesium containing a dense uniform dispersion of nanoparticles. Nature 528(7583):539–543. https://doi.org/10.1038/nature16445
7. Pan S, Sokoluk M, Cao C, Guan Z, Li X (2019) Facile fabrication and enhanced properties of Cu-40 wt% Zn/WC nanocomposite. J Alloys Compd. https://doi.org/10.1016/j.jallcom.2019.01.022
8. Sokoluk M, Cao C, Pan S, Li X (2019) Nanoparticle-enabled phase control for Arc welding of Unweldable Aluminum Alloy 7075. Nat Commun 10(1):98. https://doi.org/10.1038/s41467-018-07989-y
9. Zeng W, Xie J, Zhou D, Fu Z, Zhang D, Lavernia EJ (2018) Bulk Cu-NbC nanocomposites with high strength and high electrical conductivity. J Alloys Compd 745:55–62. https://doi.org/10.1016/j.jallcom.2018.02.215

10. Xu P, Jiang F, Tang Z, Yan N, Jiang J, Xu X, Peng Y (2019) Coarsening of Al_3Sc precipitates in Al–Mg–Sc alloys. J Alloys Compd 781:209–215. https://doi.org/10.1016/j.jallcom.2018.12.133
11. Zhang Y, Li X (2017) Bioinspired, Graphene/Al2O3 doubly reinforced aluminum composites with high strength and toughness. Nano Lett 17(11):6907–6915. https://doi.org/10.1021/acs.nanolett.7b03308
12. Yang H, Gao T, Wu Y, Zhang H, Nie J, Liu X (2018) Microstructure and mechanical properties at both room and high temperature of in-situ TiC reinforced Al–4.5 Cu matrix nanocomposite. J Alloys Compd. https://doi.org/10.1016/j.jallcom.2018.07.045
13. Pan S, Guan Z, Yao G, Cao C, Li X (2019) Study on electrical behaviour of copper and its alloys containing dispersed nanoparticles. Curr Appl Phys. https://doi.org/10.1016/j.cap.2019.01.016
14. Chang S-Y, Chen C-F, Lin S-J, Kattamis TZ (2003) Electrical resistivity of metal matrix composites. Acta Mater 51(20):6291–6302. https://doi.org/10.1016/S1359-6454(03)00462-2
15. Jiang S, Wang R (2019) Grain size-dependent Mg/Si ratio effect on the microstructure and mechanical/electrical properties of Al-Mg–Si–Sc Alloys. J Mater Sci Technol 35(7):1354–1363. https://doi.org/10.1016/j.jmst.2019.03.011
16. Roy RK, Das S (2006) New combination of polishing and etching technique for revealing grain structure of an annealed aluminum (AA1235) Alloy. J Mater Sci 41(1):289–292. https://doi.org/10.1007/s10853-005-3304-x
17. Gasparov VA, Sidorov NS, Zver'kova II, Kulakov MP (2001) Electron transport in Diborides: observation of superconductivity in ZrB2. J Exp Theor Phys Lett 73(10):532–535. https://doi.org/10.1134/1.1387521
18. Ahadi K, Shoron OF, Marshall PB, Mikheev E, Stemmer S (2017) Electric field effect near the metal-insulator transition of a two-dimensional electron system in $SrTiO_3$. Appl Phys Lett 110(6):062104. https://doi.org/10.1063/1.4975806
19. Justin JF, Jankowiak A (2011) Ultra high temperature ceramics: densification. Prop Thermal Stab AerospaceLab 3:1–11
20. Qian LH, Lu QH, Kong WJ, Lu K (2004) Electrical resistivity of fully-relaxed grain boundaries in nanocrystalline Cu. Scr Mater 50(11):1407–1411. https://doi.org/10.1016/j.scriptamat.2004.02.026
21. Basinski ZS, Dugdale JS, Howie A (1963) The electrical resistivity of dislocations. Philos Mag J Theor Exp Appl Phys 8(96):1989–1997. https://doi.org/10.1080/14786436308209092
22. Mosleh-Shirazi S, Hua G, Akhlaghi F, Yan X, Li D (2015) Interfacial valence electron localization and the corrosion resistance of Al–SiC nanocomposite. Sci Rep 5:18154. https://doi.org/10.1038/srep18154

Bioresorbable Nano-Hydroxyapatite Reinforced Magnesium Alloplastic Bone Substitute for Biomedical Applications: A Study

Somasundaram Prasadh, Vyasaraj Manakari, Gururaj Parande, T. S. Srivatsan, Raymond Wong and Manoj Gupta

Abstract Biodegradable magnesium, when reinforced with nanoparticles, displays significant improvements in its mechanical strength, cell proliferation, and corrosion resistance when compared to existing commercially available alloys. This inimitable behavior of magnesium-based composites makes it a unique offering in orthopedic and mandibular reconstruction applications. Additionally, it displays mechanical properties similar to the natural bone highlighting its huge potential as an orthopedic implant material. The present study focuses on the influence of hydroxyapatite (HA) nanoparticles on mechanical, immersion and cytocompatibility properties of magnesium processed by powder metallurgy technique coupled with hybrid microwave sintering. The inclusion of HA nanoreinforcement restricted the growth of grain size for Mg, thereby resulting in superior biodegradation and biocompatibility properties. Mg–HA nanocomposites displayed excellent corrosion resistance compared to its matrix counterpart showing a near-uniform degradation rate. MC3T3-E1 cells showed an increased cell viability percentage and subsequent low cytotoxicity

S. Prasadh · R. Wong (✉)
Faculty of Dentistry, National University of Singapore, 1 Lower Kent Ridge Road, Singapore 119083, Singapore
e-mail: denrwcw@nus.edu.sg

S. Prasadh
e-mail: e0204949@u.nus.edu

V. Manakari · G. Parande · M. Gupta
Department of Mechanical Engineering, National University of Singapore, 9 Engineering Drive 1, Singapore 117576, Singapore
e-mail: mbvyasaraj@u.nus.edu

G. Parande
e-mail: gururaj.parande@u.nus.edu

M. Gupta
e-mail: mpegm@nus.edu.sg

T. S. Srivatsan
Department of Mechanical Engineering, The University of Akron, Akron, OH 44325-3903, USA
e-mail: srivatsants@yahoo.com

T. S. Srivatsan and M. Gupta (eds.), *Nanocomposites VI: Nanoscience and Nanotechnology in Advanced Composites*, The Minerals, Metals & Materials Series, https://doi.org/10.1007/978-3-030-35790-0_6

levels. No obvious toxic effects were observed, which is consistent with the enhanced corrosion resistance of Mg alloys resulting in better cell attachment and viability. Structure–property correlations are drawn and analyzed.

Keywords Magnesium matrix composites · Corrosion · Biodegradability · Cytocompatibility · Orthopedic and mandibular reconstruction

Introduction

Metallic biomaterials have long found numerous applications in the biomedical industry owing to their excellent mechanical properties, corrosion resistance, and reasonable biocompatibility. These materials have been extensive as either permanent implants (hip, knee, elbow, spinal disc replacement) or as temporary implants like surgical screws, plates, pins, and clips in orthopedic fracture management. The most commonly used metallic biomaterials are stainless steel, titanium, and Co–Cr alloys [1–4]. These metallic biomaterials are the gold standard for permanent fixtures but come with a few drawbacks in temporary fracture management, namely, stress shielding and non-bioresorbability. The Young's modulus of the natural bone is 2–20 GPa. The metallic biomaterials like stainless steels (210 GPa), Co–Cr alloys (210 GPa), and titanium alloys (100 GPa) display a relatively higher Young's modulus values resulting in stress-shielding effects [5]. Hence, due to the mismatch in Young's modulus, during the implantation, the majority of the load is carried by the stiffer metallic implants rather than being shared by the implant and the bone, ultimately resulting in stress shielding of the surrounding bone [2, 6–8]. Literature dictates the periprosthetic bone loss to have a multifactorial etiology that includes two main causes: (i) stress shielding of bone that can trigger bone resorption as a result of changes in its loading pattern after the implantation of the stiffer implant material in the host bone tissue, eventually leading to complete bone fracture or implant loosening, and (ii) osteolysis occurring due to accumulation of debris in the host bone tissues, caused by excessive wear of implanted materials which may impede the new bone formation [9–11]. In terms of bioresorbability, none of the commercially used metallic implants is bioresorbable. These materials stay in the body for a long time even after the bone has regrown. In most cases, revision surgery is required to remove this implant from the human body. In certain cases, the implant is in the human body forever. This approach not only induces severe physical and financial trauma but also impedes the intrinsic mobility of the patient.

Biodegradable metals such as magnesium (Mg), Zinc (Zn), and iron (Fe) provide necessary structural support to host tissues, degrade gradually in vivo, and then completely dissolve after fulfilling the missions of assisting tissue healing [4, 11]. Furthermore, owing to their non-toxicity and presence in the human body as essential trace elements for performing various biological functions, Mg−, Fe−, and Zn-based alloys and composites are considered for a variety of biomedical applications. Magnesium, therefore, is a suitable replacement material considering the salient advantages it possesses over the incumbent metallic implant players in the market. Magnesium (Mg) is the lightest structural metal and is the sixth most abundant element in the earth's crust. Also, it is the fourth most abundant cation in the body responsible for close to 300 enzymatic reactions [12]. However, the Achilles heel of the magnesium research community is controlling its rapid degradation in physiological environments. In comparison to Mg, Zn does not generate gases on biodegradation and displays superior corrosion resistance and excellent processability. Nevertheless, the poor mechanical strength of pure Zn impedes its utilization as biodegradable implant material [5]. On the other hand, when compared to Mg alloys, Fe-based materials possess increased strength and superior mechanical properties but display a significantly slower degradation in the physiological environment [1, 4, 9]. Additionally, the ferromagnetic characteristics of Fe-based materials adversely affect their compatibility for magnetic resonance imaging (MRI). The elastic modulus of Mg-based materials ranges from 40–45 GPa, closer to that of natural bone (10–30 GPa) as compared to biodegradable Zn-based alloys (90–100 GPa) and Fe-based alloys (200 GPa) [5, 13]. Considering the aforementioned advantages, this class of metallic biomaterials was recognized as 'revolutionizing metallic biomaterials' for medical devices and tissue engineering [14]. Over the past two decades, various attempts have been made to control the degradation profile of magnesium-based materials in the physiological environment of the human body. One of the commonly used techniques is alloying which has proven to be effective in improving the mechanical properties and controlling the corrosion rate. However, secondary phase formation in the Mg-X alloy systems can aid in the faster degradation of magnesium. Additionally, coating technology has also been preferred by materials scientists to address the ability of controlled degradation [3, 15, 16]. However, in case of a poorly coated magnesium, the degradation is observably higher due to the galvanic corrosion. Hence, addition of inexpensive biocompatible nanoreinforcements is a cost-effective, viable option to alleviate these problems to obtain better corrosion and cytocompatibility results. Several studies have been done using secondary nanoreinforcements and its effect on the degradation and corrosion behavior of magnesium [17]. It was further understood that better dispersion of nanoscale particles in Mg matrices led to strengthening of the Mg matrices due to their inhibition of dislocation motions. For instance, the yield strength of the Mg alloy was seen to be enhanced with the addition of alumina (Al_2O_3) nanoparticles as reinforcements in AZ31B alloy [18]. Furthermore, the mechanical properties of Mg matrices via synergistic strengthening mechanisms were greatly improved by the addition of numerous nanoreinforcements in Mg matrices, such as alumina (Al_2O_3), zirconia (ZrO_2), yttria (Y_2O_3) [19], graphene nanoplatelets (GNPs) [20], and calcium phosphate ceramic (CPC) [21, 22]. The enhanced property with the presence of bioceramic nanoreinforcements like CPC is encouraging and paves way for more studies to be conducted in this domain. As bone is a bionanocomposite of collagen and hydroxyapatite, using hydroxyapatite as reinforcement in magnesium can be an interesting approach.

Accordingly, this paper extensively reviews the combined effect of HA nanoreinforcement particles along with the Mg matrix on the microstructure, biodegradability, and biocompatibility of MMNCs in the physiological environment. The effect of refined microstructure on the corrosion resistance and biocompatibility of Mg–HA nanocomposites is discussed.

Experimental Procedures

Materials and Synthesis

Pure magnesium and Magnesium–hydroxyapatite nanocomposite were synthesized using the powder metallurgy technique followed by a hybrid microwave sintering technique. This procedure is detailed in the following article [23]. The homogenized billets were soaked at 400 °C for 1 h and then hot extruded at 350 °C to obtain cylindrical rods of 8 mm. Samples cut from the rods were then characterized for microstructural, corrosion and biocompatibility response.

Microstructure

Cylindrical samples having a diameter of 8 mm and length of 5 mm were ground, finely polished, etched and observed using a LEICA-DM 2500 M metallographic light microscope (Singapore). Three representative images were analyzed for each composition to ensure high statistical conformance. The grain sizes were measured using the standard ASTM technique.

Corrosion Properties

Cylindrical samples of (5 mm diameter and 5 mm length) were immersed for 7 days in Hank's balanced salt solution (HBSS) procured from Lonza Chemicals Pte Ltd. Singapore. The setup was immersed in a water bath that was maintained at 37 °C to simulate the temperature of the human body. The solution to the sample ratio was maintained at 20 mL:1 cm^2. Weight loss and pH measurements were measured after every 24 h. A solution containing 20 g CrO_3 and 1.9 g $AgNO_3$ dissolved in 100 mL of de-ionized water was used for removing the corrosion products.

Biocompatibility

For the direct assay, 5 mm × 2 mm discs were used with direct seeding in 96 well plates. MTS solutions (CellTiter 96® Aqueous Assay System from Promega) were used for the cell proliferation assay. Direct seeding of MC3T3-E1 (8000 cells) on the discs in 96 well plate with an incubation period for 1, 3 and 5 days was performed. Post incubation, the samples with the cells were washed with PBS and 100 μl of alpha MEM media was added to the wells. 20 μl of MTS reagent was added to the wells under a dark environment and incubated for 2 h at 37 °C in 5% CO_2. Thereafter, the well plate was read for optical density (OD) values in 96 well plate reader at 490 nm wavelength ($p < 0.05$; $n = 4$). OD values obtained from the well plate reader were plotted, and the percentage of viable cells was calculated using the following formula (Eq. 1):

$$\text{Percentage of viable Cells} = \text{Experimental (OD) Values/Control (OD) values} \times 100 \quad (1)$$

The lactate dehydrogenase (LDH) activity was used as an index of the cytotoxicity in the culture media. Pure cell culture medium with cells was used for maximum LDH release with CytoTox-ONE Reagent (Promega) used as the assay. The working solution was prepared by adding 11 ml of assay buffer to 1 vial of substrate powder. MC3T3-E1 (8000 cells) were seeded directly on the discs in 96 well plates and incubated for 24 h at 37 °C in 5% CO_2. Post incubation, the plates were allowed to come to room temperature and 2 μl of lysis solution was added to the well containing the cells without the sample discs and used for maximum LDH release. 100 μl of CytoTox-ONE reagent was added to all the wells containing the sample discs. The plates were allowed to incubate at room temperature for 15 min, and the fluorescence values were recorded with excitation of 560 nm and an emission wavelength of 590 nm. ($p < 0.05$; $n = 4$). The obtained values were used to calculate the percentage of cytotoxicity using following formula (Eq. 2):

$$\text{Percentage of Cytotoxicity} = \text{Experimental Data/Maximum LDH release} \times 100 \quad (2)$$

Live–dead cell staining on the surfaces of the samples was performed using a working solution containing 100 μl propidium iodide (PI) (Dead cells) stock solution, 5 μl fluorescein diacetate (FDA) (green) (5 mg in 1 ml acetone) and 5 μl Hoechst (Blue) in 5 ml alpha MEM media without FBS and P/S. The working solution (150 μl) was added to the samples placed in 96 well plates. After incubation for 15 min at room temperature, samples were rinsed in PBS and were immediately examined using an upright fluorescence microscope (Leica DMRB, Leitz).

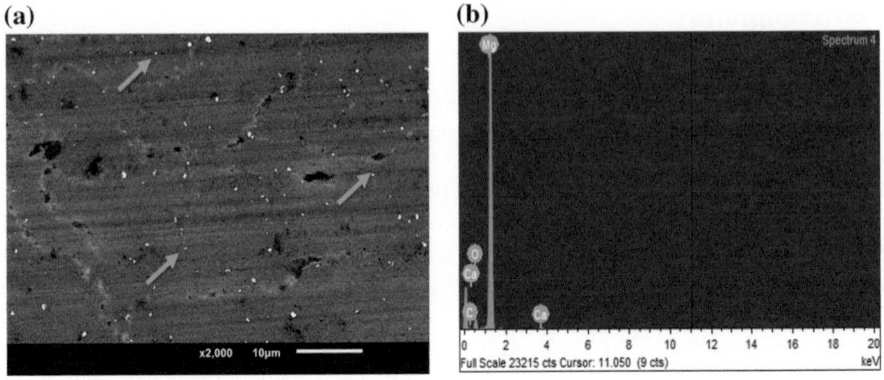

Fig. 1 SEM micrographs of **a** HA distribution and **b** EDS mapping in Mg–0.5 HA nanocomposite

Results and Discussion

Microstructure

Visual observation of the microwave sintered nanocomposite billet and the extruded rods clearly revealed an absence of any macroscopic surface defects. This confirms the overall suitability of the processing parameters used during the processes of sintering and extrusion. The intrinsic microstructural features of the developed pure Mg and Mg–0.5 HA nanocomposite were studied in terms of the presence and distribution of the reinforcing HA nanoparticles through the microstructure. Figure 1 shows the results of the Mg–0.5 HA nanocomposite. It is observed that the HA particles are homogeneously distributed through the magnesium matrix. Further, the extrusion process provided for a suitable adhesion between the magnesium particles while concurrently enabling the HA nanoparticles to be distributed homogeneously through the magnesium metal matrix, which made the secondary process much more effective in engineering a near-uniform composite.

Corrosion Properties

In order to target magnesium-based nanocomposites as a bioresorbable implant, high corrosion resistance is crucial to retain the load-bearing strength with minimum inflammatory behavior. The biocorrosion evaluation of the samples was conducted in Hank's balanced salt solution (HBSS). The HBSS was regularly changed to keep the pH equivalent to the body fluid. The corrosion and pH measurements are shown in Fig. 2. The pH values for both pure Mg and Mg–0.5 HA nanocomposites increased drastically at the end of 24 h from 7.4 with the pH values being in the range of ~9.2–9.4. The high initial pH increase can be attributed to the interaction of Mg and

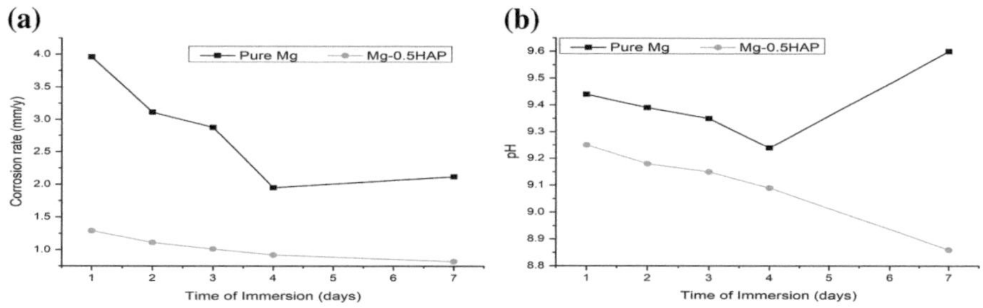

Fig. 2 Corrosion behavior of pure Mg and Mg–HA nanocomposite: **a** Calculated corrosion rates based on weight loss, and **b** pH measurements from immersion testing

Mg-based materials with physiological environments, wherein high interaction is observed for the initial 12–24 h [12, 24]. This interaction leads to the anodic dissolution of Mg and Mg-based nanocomposites releasing Mg^{2+} ions into the solution. The higher the rate of dissolution, the greater is the amount of Mg^{2+} released and higher pH values are observed. However, pH values from day 2 to day 7 for both pure Mg and Mg–0.5 HA nanocomposites reveal steady pH values, and no substantial increase or decrease in the values was observed. The absolute pH values of the nanocomposite were observed to be lesser than that of pure Mg. This suggests that the uniform presence of the HA nanoparticles has resulted in enhanced dynamic passivation of the composite samples from day 2 to day 7, thereby keeping the pH increase/decrease in control.

The corrosion rates were calculated as per Eq. 4, where, time conversion coefficient, $K = 8.76 \times 10^4$, W is the change in weight pre- and post-immersion (g), A is the surface area of the cylinder exposed to the immersive medium (cm^2), T is the time of immersion (h), and D is the experimental density of the material ($g{\cdot}cm^{-3}$) [12].

$$\text{Corrosion rate} = \frac{K \times W}{(A \times T \times D)} \quad (3)$$

The corrosion rate at the end of day 1 for pure Mg was ~3.9 mm/y. In comparison, the Mg–0.5 HA nanocomposite displayed a significantly lower corrosion rate value of 1.29 mm/y (~67% improvement). The corrosion rates of the samples decreased progressively for both pure Mg and Mg–0.5 HA nanocomposites. Mg–0.5 HA nanocomposite displayed the best response with near-uniform corrosion rate from day 2 to day 7. This improvement in the corrosion resistance can be attributed to (a) smaller grain size and (b) formation of a protective film. As seen earlier, significant grain refinement for the nanocomposite would enable the grain boundaries to act as corrosion barriers, hence increasing the corrosion resistance [12]. Note that the immersion process of the samples is governed by the ionic interactions among Mg^{2+}, Cl^-, and OH^- present in the HBSS as shown in equations [25].

$$Mg \rightarrow Mg^{2+} + 2e^- \text{ Anodic reaction} \tag{4a}$$

$$2H_2O + 2e^- \rightarrow H_2 \uparrow + 2OH^- \text{ Cathodic reaction} \tag{4b}$$

$$Mg^{2+} + 2OH^- \rightarrow Mg(OH)_2 \tag{4c}$$

$$Mg(OH)_2 + Cl^- \rightarrow MgCl_2 + 2OH^- \text{Chloride attack} \tag{4d}$$

$$10Ca^{2+} + 8OH^- + 6HPO_4^{2-} \rightarrow Ca_{10}(PO_4)_6 + 6H_2O \text{ Apatite biomineralization} \tag{4e}$$

$$10Ca^{2+} + 2OH^- + 6HPO_4^{2-} \rightarrow Ca_{10}(PO_4)_6$$

$$3Ca^{2+} + 2OH^- + 2PO_4^{3-} \rightarrow (Ca)_3(PO_4)_2$$

Anodic dissolution of Mg into Mg^{2+} results in the increased corrosion rate and pH values from day 0 to day 1. This anodic dissolution leads to the formation of hydrogen gas pockets at the surface of the sample. pH of the solution and hydrogen gas evolution has a close correlation and can provide a near accurate estimation of the extent of gas pockets formed [26]. Further, a protective magnesium hydroxide layer is formed due to the ionic interaction between Mg^{2+} and OH^-. This results in a decrease in the corrosion rate as the hydroxide layer forms a diffusion barrier between the matrix and the solution. The low radius Cl^- ions diffuse through the layer and interacts with the surface resulting in localized pits and with the secondary phases causing micro-galvanic corrosion [8]. Additionally, the nanocomposite degradation releases Ca^{2+}, $PO_4{}^{3-}$, and $HPO_4{}^{2-}$ ions which react with the free OH^- to form $(Ca)_3(PO_4)_2$ apatite layer. The primary reason for the growth of $Ca_{10}(PO_4)_6$ and $(Ca)_3(PO_4)_2$ is the degree of supersaturation in the physiological environment. HA, when compared to other calcium phosphates, is the most stable with the lowest solubility [27]. Thus, the corrosion products formed are a combination of Mg $(OH)_2$, MgO and $(Ca)_3(PO_4)_2$ compounds. To note that high pH and Mg^{2+} concentration can damage the cell viability in a culture medium and therefore formation of the corrosion products assists in cell attachment to the samples. The current study indicates that the presence of HA nanoparticles can improve the biocompatibility of pure Mg [28].

Biocompatibility

The absorbance values are directly and precisely proportional to the amount of cell proliferation. Figure 3a shows the results of MTS assay of pure Mg and Mg–0.5 HA nanocomposite by MC3T3-E1 pre-osteoblast-like cells. The cell viability of MC3T3-E1 pre-osteoblast cells expressed as a percentage of the viability of cells cultured in the negative control after incubation for 1, 3 and 5 days of the samples. As it is shown,

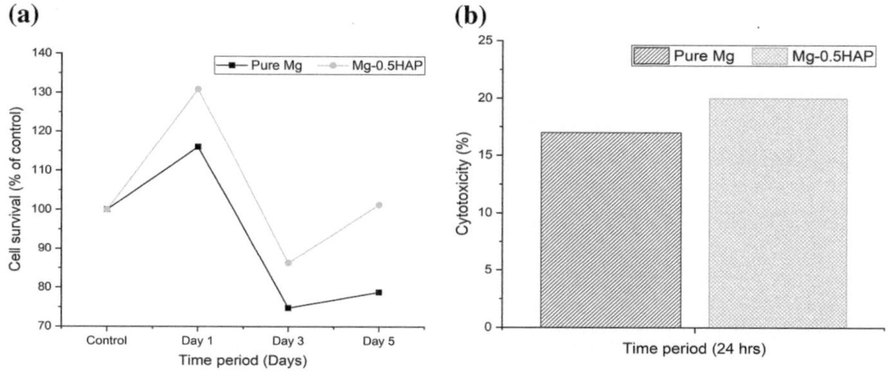

Fig. 3 **a** Cell survival in percentage with respect to time, and **b** Cytotoxicity measurement using LDH release assay for the developed materials

the pre-osteoblast cells can attach and proliferate on both the samples but there was an increase in cell proliferation recorded for Mg–0.5% HA nanocomposites. Mg–0.5 HA nanocomposite samples improved and accelerated cell proliferation compared to the cell growth on pure Mg after day 1, 3 and day 5 incubation. Nevertheless, after 3 and 5 days, cell proliferation on Mg–0.5 HA nanocomposites was considerably ($P < 0.05$) higher in comparison to pure Mg. As per the ISO 10993–5:2009 standard for MTT assay, if the cell viability percentage is greater than 70% of the negative control, the material has zero cytotoxicity potential. This suggests that both the pure Mg and HA nanocomposites exhibit satisfactory biosafety with regard to cellular metabolism. This cytocompatibility response is consistent with the low corrosion rates of the samples aiding in superior cell attachment and proliferation and high cell survivability.

Cytotoxicity, indicated by the lactate dehydrogenase (LDH) activity in the culture media after 24 h of incubation, was analyzed in MC3T3E1 cells cultured as shown in Fig. 3b and was calculated by normalizing to maximum lysis (100% toxicity) and spontaneous LDH release or 0% toxicity controls. Cytotoxicity values falling within this specified range are considered nontoxic. The amount of LDH released from the dead cells when lysed with a diluted detergent is corresponding to the maximum lysis control. LDH assay results (Fig. 3b) revealed a level 0 cytotoxicity ($\leq$25% cell death) for both pure Mg and Mg–0.5 HA nanocomposites samples after 24 h of exposure. HA nanocomposite sample showed higher cytotoxicity compared to pure Mg but it is within the maximum LDH release concentration of 100%. No latent cytotoxic effects were observed in the current study.

Cell viability was evaluated using live and dead cell staining (Fig. 4). The results show more live cells (green) and nucleus of live cells (blue) for Mg–HA nanocomposite compared to pure Mg. Cell proliferation and cell densities increased from day 1 to day 5 for Mg–HA nanocomposite compared to pure Mg. Dead cells(red) were more in pure Mg on day 3 incubation. Less density of dead cells was found for Mg–HA nanocomposite samples. The cell distribution was overall better for the Mg–HA nanocomposite as shown in Fig. 4.

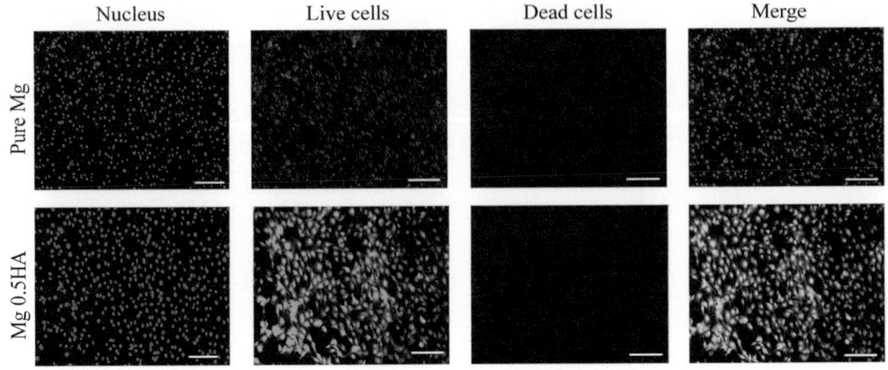

Fig. 4 Representative fluorescent microscopy images showing live/dead staining of primary MC3T3-E1 pre-osteoblasts cultured on pure Mg and Mg–HA nanocomposite samples for day 3 (10 × magnification and the scale bar indicate 200 μm)

Conclusions

Motivated by the versatile properties of magnesium and its potential in orthopedic and mandibular reconstruction, magnesium-based hydroxyapatite nanocomposites were synthesized in this study using powder metallurgy technique followed by hybrid microwave sintering. The presence of hydroxyapatite assisted in superior refinement of the microstructural features of magnesium. Enhanced corrosion protection and a near-uniform biodegradation rate were observed for the nanocomposite when compared to the monolithic samples. Better cell viability was observed for the nanocomposite, and little to no change was observed in the cytotoxicity values of both the samples. However, the nanocomposite showed slightly higher cytotoxic values, still within level 0 cytotoxicity levels. Better cell attachment and proliferation were observed in the nanocomposite owing to the combined effect of refined microstructure, improved corrosion resistance, and better cell viability, in comparison with monolithic magnesium.

References

1. Chen Q, Thouas GA (2015) Metallic implant biomaterials. Materials science and engineering: R: reports 87:1–57
2. Shayesteh Moghaddam N, Taheri Andani M, Amerinatanzi A, Haberland C, Huff S, Miller M, Elahinia M, Dean D (2016) Metals for bone implants: safety, design, and efficacy. Biomanufacturing Rev 1:1
3. Durairaj R, Ramachandran S (2018) In vitro characterization of electrodeposited hydroxyapatite coatings on titanium (Ti6AL4 V) and magnesium (AZ31) alloys for biomedical application. Int J Electrochem Sci 13:4841–4854
4. Eliaz N (2019) Corrosion of metallic biomaterials: a review. Materials (Basel, Switzerland) 12:407

5. Gupta M, Meenashisundaram GK (2015) Insight into designing biocompatible magnesium alloys and composites: processing, mechanical and corrosion characteristics. Springer
6. Tekumalla S, Yang C, Seetharaman S, Wong WLE, Goh CS, Shabadi R, Gupta M (2016) Enhancing overall static/dynamic/damping/ignition response of magnesium through the addition of lower amounts (<2%) of yttrium. J Alloy Compd 689:350–358
7. Shi L, Shi L, Wang L, Duan Y, Lei W, Wang Z, Li J, Fan X, Li X, Li S, Guo Z (2013) The improved biological performance of a novel low elastic modulus implant. PLoS ONE 8:e55015–e55015
8. Nayak S, Bhushan B, Jayaganthan R, Gopinath P, Agarwal RD, Lahiri D (2016) Strengthening of Mg based alloy through grain refinement for orthopaedic application. J Mech Behav Biomed Mater 59:57–70
9. Katti KS (2004) Biomaterials in total joint replacement. Colloids Surf B Biointerfaces 39:133–142
10. Staiger MP, Pietak AM, Huadmai J, Dias G (2006) Magnesium and its alloys as orthopedic biomaterials: a review. Biomaterials 27:1728–1734
11. Zheng YF, Gu XN, Witte F (2014) Biodegradable metals. Mater Sci Eng: R: Reports 77:1–34
12. Parande G, Manakari V, Gupta H, Gupta M (2018) Magnesium-β-Tricalcium phosphate composites as a potential orthopedic implant: a mechanical/damping/immersion perspective. Metals 8:343
13. Manakari V, Parande G, Gupta M (2017) Selective laser melting of magnesium and magnesium alloy powders: a review. Metals 7:2
14. Yun Y, Dong Z, Lee N, Liu Y, Xue D, Guo X, Kuhlmann J, Doepke A, Halsall HB, Heineman W (2009) Revolutionizing biodegradable metals. Mater Today 12:22–32
15. Tang H, Tao W, Wang C, Yu H (2018) Fabrication of hydroxyapatite coatings on AZ31 Mg alloy by micro-arc oxidation coupled with sol–gel treatment. RSC Advances 8:12368–12375
16. Tian Q, Lin J, Rivera-Castaneda L, Tsanhani A, Dunn ZS, Rodriguez A, Aslani A, Liu H (2019) Nano-to-submicron hydroxyapatite coatings for magnesium-based bioresorbable implants—deposition. Charact Degrad Mech Prop Cytocompatibility Sci Reports 9:810
17. Shahin M, Munir K, Wen C, Li Y (2019) Magnesium matrix nanocomposites for orthopedic applications: a review from mechanical, corrosion, and biological perspectives. Acta Biomater 96:1–19
18. Nguyen Q, Gupta M (2008) Enhancing compressive response of AZ31B magnesium alloy using alumina nanoparticulates. Compos Sci Technol 68:2185–2192
19. Sanaty-Zadeh A (2012) Comparison between current models for the strength of particulate-reinforced metal matrix nanocomposites with emphasis on consideration of Hall-Petch effect. Mater Sci Eng, A 531:112–118
20. Patelis N, Moris D, Matheiken S, Klonaris C (2016) The potential role of graphene in developing the next generation of endomaterials. BioMed Res Int 2016(Article ID 3180954): 7
21. Xu HH, Sun L, Weir MD, Takagi S, Chow LC, Hockey B (2007) Effects of incorporating nanosized calcium phosphate particles on properties of whisker-reinforced dental composites. J Biomed Mater Res Part B: Appl Biomater: Off J Soc Biomater Jpn Soc Biomater Aust Soc Biomater Korean Soc Biomater 81:116–125
22. Roeder RK, Turner CH (2010) Composite biomaterial including anisometric calcium phosphate reinforcement particles and related methods, in, Google Patents
23. Kujur MS, Manakari V, Parande G, Tun KS, Mallick A, Gupta M (2018) Enhancement of thermal, mechanical, ignition and damping response of magnesium using nano-ceria particles. Ceram Int 44(13):15035–15043
24. Beck GR Jr, Ha S-W, Camalier CE, Yamaguchi M, Li Y, Lee J-K, Weitzmann MN (2012) Bioactive silica-based nanoparticles stimulate bone-forming osteoblasts, suppress bone-resorbing osteoclasts, and enhance bone mineral density in vivo, Nanomedicine: Nanotechnology. Biol Med 8:793–803
25. Bakhsheshi-Rad HR, Idris MH, Abdul-Kadir MR, Ourdjini A, Medraj M, Daroonparvar M, Hamzah E (2014) Mechanical and bio-corrosion properties of quaternary Mg–Ca–Mn–Zn alloys compared with binary Mg–Ca alloys. Mater Des 53:283–292

26. Wong P-C, Tsai P-H, Li T-H, Cheng C-K, Jang JSC, Huang JC (2017) Degradation behavior and mechanical strength of Mg-Zn-Ca bulk metallic glass composites with Ti particles as biodegradable materials. J Alloy Compd 699:914–920
27. Del Campo R, Savoini B, Muñoz A, Monge M, Garcés G (2014) Mechanical properties and corrosion behavior of Mg–HAP composites. J Mech Behav Biomed Mater 39:238–246
28. Razavi M, Fathi M, Savabi O, Tayebi L, Vashaee D (2018) Improvement of in vitro behavior of an Mg alloy using a nanostructured composite bioceramic coating. J Mater Sci Mater Med 29:159

Part III
Processing–Properties–Performance of Nanocomposites

Investigating and Understanding the Mechanical and Tribological Properties of a Magnesium Hybrid Metal–Ceramic Nanocomposite

R. Arvind Singh, S. Jayalakshmi, S. Sankaranarayanan, Xizhang Chen, S. Konovalov, T. S. Srivatsan and M. Gupta

Abstract A major limitation of magnesium-based materials is their low wear resistance, which results in premature failure of components during service. In this paper, the tribological behaviour of Mg–5Nb metal–metal composite reinforced with varying weight fractions of nano-sized silicon carbide nano-silicon carbide (SiCn) ceramic particles is presented and briefly discussed. The composites were made using the technique of disintegrated melt deposition (DMD) followed by hot extrusion. Tribological test results showed improved wear resistance of the nanocomposites, which can be attributed to the conjoint and mutually interactive influences of grain refinement, presence of hard, brittle and elastically deforming SiCn particles, and

R. Arvind Singh (✉) · S. Jayalakshmi · X. Chen · S. Konovalov
College of Mechanical and Electrical Engineering, Wenzhou University, Wenzhou 325035, Zhejiang, China
e-mail: arvindsingh.r@gmail.com

S. Jayalakshmi
e-mail: jayalakshmi.subramanian@gmail.com

X. Chen
e-mail: chenxizhang@wzu.edu.cn

S. Konovalov
e-mail: ksv@ssau.ru

S. Sankaranarayanan
ANSYS Software Private Ltd, Bangalore 560034, Karnataka, India
e-mail: ssnseetharaman@yahoo.com

S. Konovalov
Institute of Rocket and Space Technology, Samara National Research University, Samara 443086, Russia

T. S. Srivatsan
Department of Mechanical Engineering, The University of Akron, Akron, OH 44325-2903, USA
e-mail: tss1@uakron.edu

M. Gupta
Department of Mechanical Engineering, National University of Singapore, 9 Engineering Drive 1, Singapore 117576, Singapore
e-mail: mpegm@nus.edu.sg

T. S. Srivatsan and M. Gupta (eds.), *Nanocomposites VI: Nanoscience and Nanotechnology in Advanced Composites*, The Minerals, Metals & Materials Series, https://doi.org/10.1007/978-3-030-35790-0_7

the formation and presence of hard and brittle intermetallic phases. These in synergism contribute to increasing the hardness of the nanocomposites with a concomitant improvement in wear resistance. From a careful and comprehensive analysis of the worn surface analysis, the governing wear mechanisms were identified for the developed materials.

Keywords Pure magnesium · Metallic reinforcement · Nano-silicon carbide · Hardness · Wear · Friction

Introduction

Magnesium alloys and their composite counterparts, due to their low density and high strength-to-weight (σ/ρ) ratio, have the potential to improve energy efficiency and system performance in industries spanning aerospace, automobile, defense, and even commercial products. In comparison with the magnesium (Mg) alloys, the magnesium (Mg)-based composites provide an improvement in hardness, elastic modulus, yield strength (σ_{YS}), ultimate tensile strength (σ_{UTS}) and elevated temperature properties. However, ductility of the composites is severely impaired by the addition and/or presence of micron-sized ceramic reinforcements in the metal matrix [1, 2]. This results in poor toughness (work of fracture) of the composites. In an attempt to improve the ductility of magnesium and its alloys, addition of nano-sized reinforcements to the Mg-based nanocomposites (Mg–MMNCs) [3, 4] is being both researched and examined. The nanocomposites reveal a noticeable improvement in ductility when compared to the micron-size reinforced composites, while concurrently possessing enhanced toughness [3, 4].

A major limitation of the magnesium-based materials is their low wear resistance. When two materials undergo relative mechanical motion, both friction and wear come into force. Wear is often an undesirable loss of material that eventually culminates in catastrophic failure. Magnesium-based materials are viable candidates for tribological applications, such as (i) brakes, (ii) brake drums, and (iii) piston ring cylinder, to name just a few. In these applications, it is important to control both the wear and friction characteristics. A significant improvement in wear resistance has been widely reported for Mg composites with micron-sized reinforcements [5, 6]. However, a major drawback of these Mg composites with micron-sized reinforcements is the occurrence of counter-abrasion by the hard, brittle and essentially elastically deforming ceramic particles [5, 6]. In this context, the tribological performance of nanocomposites is of both scientific and technological interests.

In this innovative research study, Mg–5Nb is reinforced with varying weight fractions of nano-silicon carbide (SiCn) ceramic particles made by the technique of disintegrated melt deposition (DMD). The engineered and/or synthesized composites were then characterized for their microstructure, hardness, and sliding wear behavior. The wear behavior and the operating wear mechanism of the nanocomposite were studied in comparison to that of the Mg–5Nb.

Experimental Details

Primary Processing

Pure magnesium was reinforced with 5 weight pct. niobium (Nb) and 0.25 weight pct. and 2.0% (in weight pct.) nano-silicon carbide particulates (SiCn). The magnesium turnings [provided by Acros Organics, Singapore], niobium powder with an average particle size <10 μm [acquired from Prochem, Rockford Illinois] and SiCn particles with an average particle size ~50 nm [acquired from Nanostructured and Amorphous, USA] were used. The technique of Disintegrated Melt Deposition (DMD), which employs bottom pouring of the melt, was used to produce the magnesium nanocomposites [3]. The DMD technique adopts a simultaneous vortex stirring of the melt coupled with its disintegration in an atmosphere of argon (Ar) gas. The argon gas was purged into the molten metal at about 3 L/min at a super-heat temperature of 750 °C, with intermittent stirring for 5 min. The melt was then gradually released through an orifice having a diameter of 10 mm, situated at the bottom of the crucible, and disintegrated using two jets of argon gas maintained at a flow rate of 25 L/min (oriented in a direction normal to the melt stream) prior to its deposition on to a steel mold.

Secondary Processing

The Mg-hybrid composites were machined to 36 mm diameter and then soaked at 400 °C for 60 min. Hot extrusion was carried out using a 150-ton hydraulic press at 350 °C and an extrusion ratio of 20.25:1 to obtain rods that measured 8 mm in diameter that were used for further characterization.

Materials Characterization

Microstructural characterization of the finely polished and etched samples were conducted using a field emission scanning electron microscope (Hitachi FESEM-S4300) coupled with an energy dispersive analysis (EDS). X-ray diffraction analyses were conducted on the samples using an automated Shimadzu LAB-XRD-6000 X-ray diffractometer (Cu Kα; $\alpha = 1.54056$ Å) using a scanning speed of 2° min^{-1} to study the formation of phases. Microhardness measurements were made on the polished flat specimens using a Matsuzawa MXT 50 automatic digital microhardness tester equipped with a Vickers indenter [load: 25 gf and dwell time: 15 s]. More than 10 measurements were made on each test specimen. The wear experiments were conducted under dry sliding conditions at ambient temperature [27° C] using a pin-on-disc machine. The experiments were conducted at a constant sliding speed of

1 m/s, with varying applied normal loads (10 N, 30 N and 50 N). The experiments were conducted against a hardened EN31 steel disc [hardness: 62 HRc, roughness: 30 μm]. Cylindrical pins of the developed magnesium materials, with a diameter of 6 mm, were used as the test specimens. The wear (in terms of loss in height) was measured using a linear variable differential transformer (LVDT), which had a range of 2000 μm with a resolution of ± 1 μm. The frictional force was measured using a load cell capable of measuring up to 20 N. The experiments were repeated three times, and the average values chosen for the purpose of graphical plots. Both height loss and friction force were plotted as a function of sliding distance. From the gathered test data, the wear rates (mm^3/m) and coefficient of friction were determined. Worn surface analysis was performed using a field emission scanning electron microscope (FESEM) [Model: Quanta 200] to help establish or identify the operating wear mechanisms.

Results and Discussion

Microstructure

Table 1 summarizes the results of the microstructural investigations conducted on the extruded magnesium materials. It can be seen that near-equiaxed grains were observed in all the materials, with an average grain size <10 μm. The refined microstructure is due to the presence of both niobium (Nb) and nano-silicon carbide (SiCn) reinforcement particles, which act as heterogeneous sites for grain nucleation [7]. Further, as the hot extrusion was carried out at 350 °C (>0.5 Tm), it facilitated in the occurrence of recrystallization, thereby resulting in the formation and presence of fine and nearly equiaxed shaped grains [3]. When compared to the Mg–5Nb, addition of 0.25% SiCn did not contribute appreciably to a reduction in grain size. However, for the 2.0 % SiCn addition resulted in an observable reduction in the grains size of the engineered composite. This provides an indication that for 2.0% SiCn content, the nanoparticles act as obstacles for grain growth [7], thereby resulting in a fine-grained structure. Microstructures of the Mg–5Nb composite (Fig. 1a) and Mg–5Nb–SiCn composites (Fig. 1b, c) are shown in Fig. 1.

Table 1 Grain size and morphology of Mg–5Nb and Mg–5Nb–SiCn composites

Material	Grain Size/μm	Aspect ratio	Roundness/μm
Mg–5Nb	9.1 ± 3.2	1.8 ± 0.63	1.9 ± 0.6
Mg–5Nb–0.25% SiCn	9.9 ± 4.4	1.8 ± 0.5	1.6 ± 0.3
Mg–5Nb–2.0% SiCn	5.9 ± 2.6	1.8 ± 0.7	1.6 ± 0.4

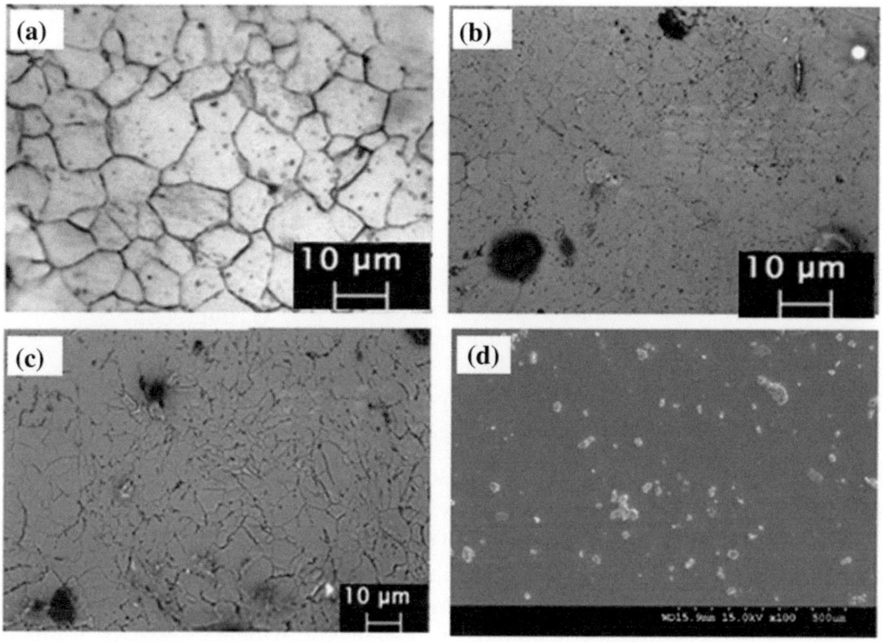

Fig. 1 Optical micrographs of the developed materials **a** Mg5Nb, **b** Mg–5Nb–0.25% SiCn, **c** Mg–5Nb–2.0% SiCn, and **d** SEM micrograph showing agglomeration of Nb particles in Mg–5Nb–0.25% SiCn

Figure 1d shows a scanning electron microscope (SEM) image of the Mg–5Nb–SiCn composite that reveals a distribution of the niobium (Nb) particles in the magnesium (Mg) metal matrix. It is seen that the Nb particles tend to form clusters and are not uniformly distributed through the metal matrix. From EDS elemental distribution (not shown here), it was identified that the reinforcing nano-sized silicon carbide (SiCn) particles are present within the Nb clusters. Particle clustering can occur due to processing parameters, such as (i) improper melt stirring time and (ii) insufficient melt super-heat temperature, or due to poor interfacial properties. Interfacial properties depend on wettability of the reinforcing particle with the matrix and bonding between the matrix and the reinforcing particles. It should be noted that based on the Mg–Nb phase diagram, Nb and Mg have no mutual solubility and do not form an intermetallic phase [8]. Hence, the Nb particles will tend to agglomerate during processing due to the relatively high density of Nb [8.57 g/cc] coupled with the absence of chemical bonding between the Nb particles and the Mg matrix. XRD analysis (Fig. 2) reveals the Mg and Nb peaks, along with very low-intensity Mg_2Si peak. However, the SiC phase is not visible, probably due to their very fine size and low weight fraction of the particles.

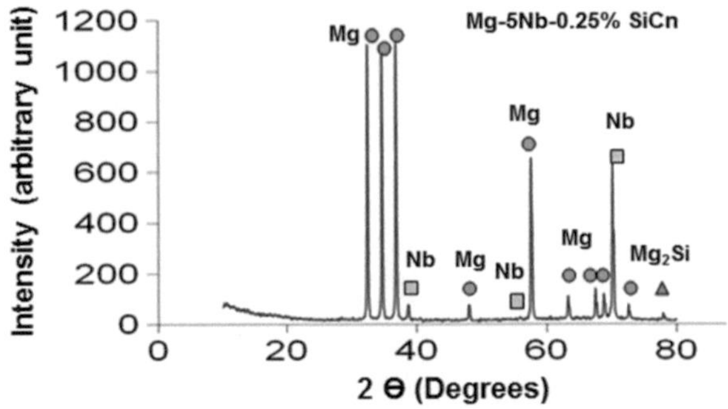

Fig. 2 Representative XRD graph of Mg–5Nb–0.25% SiCn composite. A very low fraction of the Mg_2Si phase can be seen. The SiC peak is not visible due to its small particle size and low weight fraction

Microhardness

The microhardness values of the developed Mg materials are summarized in Fig. 3. The observed increase in microhardness values for both the composites when compared to the Mg–5Nb can be attributed to the presence of relatively hard reinforcements in the metal matrix [3]. Among the composites, the Mg–5Nb–2.0% SiCn composite showed the highest hardness value that was twice higher than the Mg–5Nb metal–metal composite. A high hardness value of the nanocomposite is due to the conjoint and mutually interactive influences of the following: (i) grain refinement, (ii) increased SiCn content, and (iii) an increase in the formation and presence of the Mg_2Si.

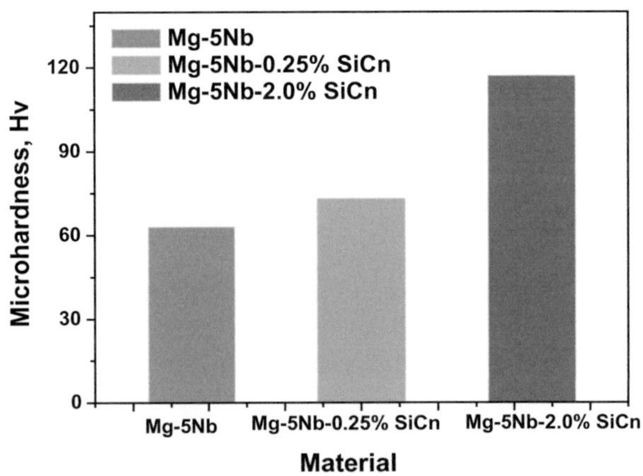

Fig. 3 Microhardness of the developed Mg–Nb–SiCn composites

Wear and Friction Behavior

The volumetric wear rate and coefficient of friction of Mg–5Nb, Mg–5Nb–0.25% SiCn and Mg–5Nb–2.0% SiCn at applied normal loads of 10 N, 30 N and 50 N are shown in Fig. 4a, b, respectively. It is seen that the wear rate increases with an increase in load for all the test samples. At all applied normal loads, the wear rate decreases with an increase in the addition of reinforcing SiC nanoparticles, except for the Mg–5Nb–0.25% SiCn composite at 50 N. The coefficient of friction decreases with an increase in applied load for all the test samples.

The nanocomposites have better wear resistance than Mg–5Nb (except for Mg–5Nb–0.25% SiCn at 50 N). Their wear behavior is consistent with the Archard's equation for wear, which states that wear of a material is inversely related to its hardness [9]. This wear equation provides a relation between volumetric wear and hardness of a material to be Q = kW/H [where Q is the volume worn per unit sliding distance, W is the applied load, H is the hardness and k is the wear coefficient] [9].

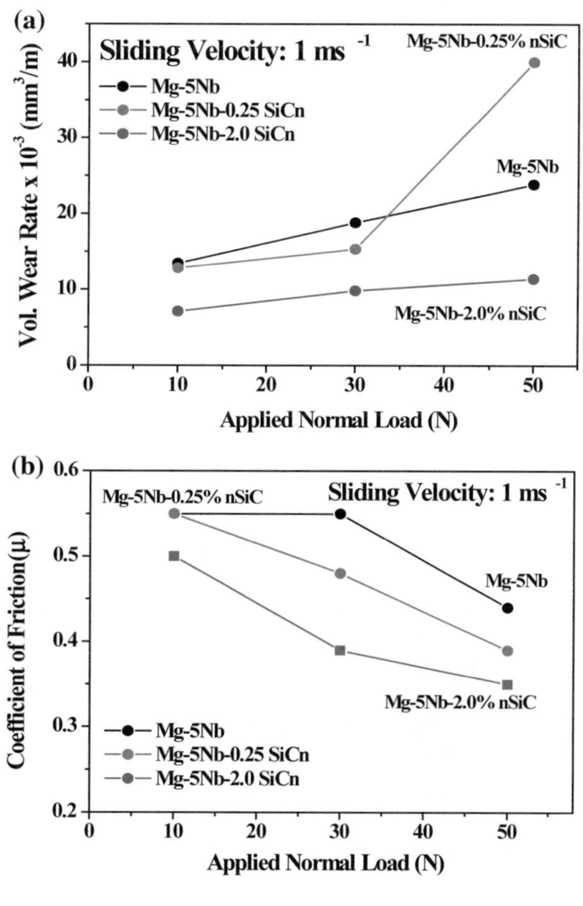

Fig. 4 **a** Volumetric wear rate, and **b** coefficient of friction of the developed materials, as a function of applied normal load

The nanocomposites showed better wear resistance than Mg–5Nb, due to their higher hardness. Among the nanocomposites, the Mg–5Nb–2.0% SiCn composite showed the best wear resistance. The high wear rate of Mg–5Nb–0.25% SiCn composite at 50 N can be attributed to a severe agglomeration of the Nb particles, which tends to lower the load-bearing capacity while concurrently promoting easy crack propagation at higher loads [10].

The coefficient of friction of the nanocomposites decreases with an increase in the addition of reinforcing SiC nanoparticles, which lowers the adhesion (i.e. stickiness) between the contacting surfaces due to the presence of nano-ceramic particles. The best tribological characteristics, i.e. low wear rate and low coefficient of friction of the Mg–2.0% SiCn composite is attributed to the following [11]:

(a) An increased weight fraction of SiCn particles,
(b) Uniform distribution of SiC and Nb particles,
(c) Presence of optimal amount of reinforcement required for improvement in wear resistance,
(d) Refined grain size, and
(e) Increased hardness.

The worn surfaces of the developed materials were observed using a SEM to both identify and establish the operating wear mechanisms. Representative SEM images of the worn surfaces of Mg–5Nb and Mg–5Nb–2.0% SiCn are shown in Figs. 5 and 6, respectively. The SEM images reveal wear mechanisms of abrasion and plastic flow for the Mg–5Nb and its nanocomposites. Further, it can be understood that

(i) At low load (10 N) abrasion occurs in both the Mg–5Nb and its nanocomposites (Figs. 5a, 6a).
(ii) At high load (50 N) plastic flow occurs in both Mg–5Nb and its nanocomposites (Figs. 5b and 6b).

The intensity of wear mechanisms, namely, abrasion and plastic flow, are severe for the Mg–5Nb when compared to the engineered nanocomposites. Therefore, wear

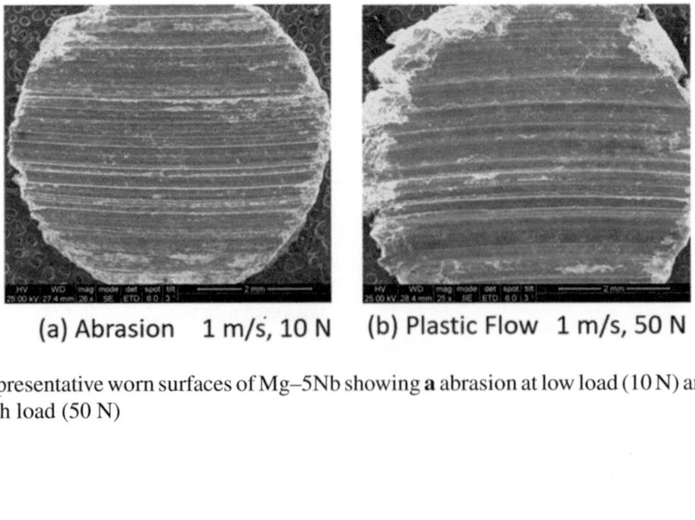

(a) Abrasion 1 m/s, 10 N (b) Plastic Flow 1 m/s, 50 N

Fig. 5 Representative worn surfaces of Mg–5Nb showing **a** abrasion at low load (10 N) and **b** plastic flow at high load (50 N)

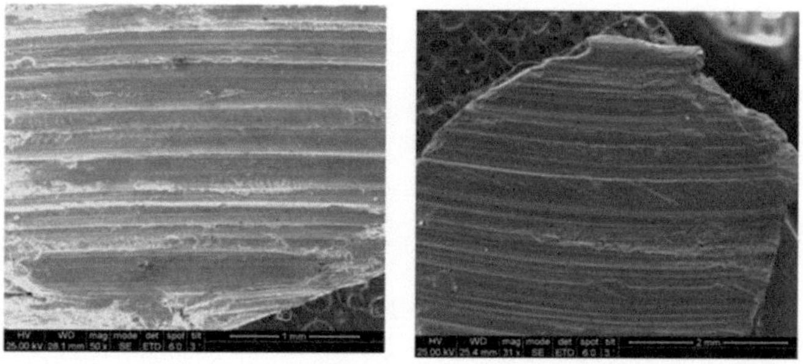

(a) Less Abrasion 1 m/s, 10 N (b) Less Plastic Flow 1 m/s, 50 N

Fig. 6 Representative worn surfaces of Mg–5Nb–2.0% SiCn showing **a** less abrasion at low load (10 N) and **b** less plastic flow at high load (50 N)

rate of the Mg–5Nb is higher than that of the nanocomposites, the exception being Mg–5Nb–0.25% SiCn at 50 N.

It is important to note that the Mg–MMCs contain micron-sized reinforcements, and third body effect is prominent. Third body effect indicates counter-abrasion of the surfaces by wear debris containing hard micro-reinforcements, which is a major drawback for the Mg–MMCs having micron-sized reinforcements [5, 6]. In contrast, nano-sized reinforcements do not induce severe counter-abrasion due essentially to their small size and low content.

Conclusions

Mg–5Nb–SiCn nanocomposites with varying SiCn weight fractions were produced by the technique of disintegrated melt deposition (DMD) followed by hot extrusion. The tribological behaviour of the nanocomposites was investigated and following are the key conclusions that can be drawn from this innovative research study.

1. Wear rate of the Mg–5Nb–SiCn-based nanocomposites increased with an increase in load, whereas their coefficient of friction decreased with an increase in load.
2. For all the test materials, at low load, wear occurs by abrasion. At high load, plastic deformation, i.e. flow, causes wear. The wear mechanisms are less intense in the nanocomposites when compared to Mg–5Nb material.
3. The content of silicon carbide nanoparticles and the uniform distribution of Nb+SiCn particles does exert an influence on the wear behavior of the nanocomposites significantly.
4. The Mg–5Nb–2SiCn showed the best wear resistance and lowest coefficient of friction. The excellent wear resistance of the composite is attributed to its high hardness caused by the conjoint influence of the following: (i) higher content of SiCn, (ii) grain refinement and (iii) presence of hard Mg_2Si formed during processing.

References

1. Chawla N, Chawla KK (2006) Metal matrix compostes. Springer, New York
2. Kainer K (2006) Metal matrix composites: custom-made materials for automotive and aerospace engineering. Wiley, UK
3. Ceschini L, Dahle A, Gupta M, Jarfors A, Jayalakshmi S, Morri A, Rotundo F, Toschi S, Arvind Singh R (2016) Aluminium and Magnesium metal matrix nanocomposites. Springer Nature, Singapore
4. Gupta M, Sharon NML (2011) Magnesium, Magnesium Alloys, and Magnesium composites. Wiley, New Jersey
5. Jayalakshmi S, Kailas SV, Seshan S, Fleury E (2006) Properties of squeeze cast Mg–6Zn–3Cu alloy and its saffil alumina short fibre reinforced composites. J Mater Sci 41:3743–3752
6. Jayalakshmi S, Kailas SV, Seshan (2003) Properties of squeeze cast Mg–10Al–Mn alloy and its alumina short fibre composites. J Mater Sci 38:1383–1389
7. Jayalakshmi S, Zhenhua L, Sankaranarayanan S, Hamouda AMS, Gupta M (2012) Microstructure and mechanical properties of Mg–5Nb metal-metal composites reinforced with nano-SiC ceramic particulates. Metals 2:178–184
8. Smith JF (1990) Mg–Nb (magnesium–niobium) binary alloy phase diagrams. In: Massalski TB (ed) ASM alloy phase diagrams center, II. ASM International. Vol 3. Materials Park, OH, USA, p 2526
9. Archard J (1953) Contact and rubbing of flat surfaces. J Appl Phys 24(8):981–988
10. Hong SJ, Kim HM, Huh, Suryanarayana D, Chun BS (2003) Effect of clustering on the mechanical properties of SiC particulate-reinforced aluminum alloy 2024 metal matrix composites. Mater Sci Eng A 347:198–204
11. Jayalakshmi S, Arvind Singh R (2015) Processing routes and tribological properties of light metal matrix nano-composites. In: Processing techniques and tribological behavior of composite materials, IGI Global, Pennsylvania, USA. ISBN: 9781466675315

Fabrication of Silver–Rhodium Nanomaterials for Chemical Sensing Applications

Simona E. Hunyadi Murph and K. J. Coopersmith

Abstract We report a novel class of silver–rhodium (Ag–Rh) nanomaterials that can be used as surface-enhanced Raman scattering (SERS) substrates. Ag–Rh nanomaterials were prepared via the co-reduction of the metal precursors by polyol at elevated temperature. Nanomaterials were characterized by scanning electron microscopy, UV–Visible spectroscopy, and energy-dispersive X-ray analysis (EDS), and they were evaluated for their ability to promote surface-enhanced Raman scattering of a model analyte.

Keywords Silver–rhodium nanoparticles · SERS · Raman

Introduction

Detection of trace quantities of analytes is critical in virtually every scientific discipline, ranging from part per billion analyses of pollutants in sub-surface water to analysis of cancer treatment drugs in blood serum [1–3]. There is a constant need in designing sensors with substantially smaller size, lower weight, faster detection, greater sensitivity, better specificity and with more modest power requirements. Sensors find applications in many industries, among them transportation, communications, medicine, safety, forensic, and national security.

Nanotechnology enables the creation of innovative and functional materials, devices, and systems by controlling matter at the atomic and molecular scales [1–6]. Nanoparticles interact strongly with light waves, even though the wavelength of the light may be much larger than the particle. They are quite different from the same materials in the bulk, which do not exhibit quantum effects. In metal nanoparticles, particularly gold and silver, "plasma oscillations" driven by external electromagnetic

S. E. Hunyadi Murph (✉) · K. J. Coopersmith
National Security Directorate, Savannah River National Laboratory, Aiken, SC, USA
e-mail: simona.murph@srnl.doe.gov

S. E. Hunyadi Murph
Department of Physics and Astronomy, University of Georgia, Athens, GA, USA

T. S. Srivatsan and M. Gupta (eds.), *Nanocomposites VI: Nanoscience and Nanotechnology in Advanced Composites*, The Minerals, Metals & Materials Series, https://doi.org/10.1007/978-3-030-35790-0_8

fields are localized and lead to strong resonances at specific wavelengths that are dependent on the particle size, shape and the local dielectric environment [1–4]. In addition, particles with sharp features are expected to behave as optical antenna by concentrating the local field into a small volume [5, 6]. Plasmonic properties of metallic nanoparticles have potential application in plasmon-enhanced spectroscopy, near-field imaging, sensing and nanophotonic devices [1–6]. By exploiting these novel properties and phenomena at the nanoscale, novel and improved detection technologies can be produced.

Over the years, a great deal of interest has been directed toward the development of active substrates for surface-enhanced Raman scattering (SERS) experiments. Surface-Enhanced Raman Spectroscopy (SERS) is a surface-sensitive analytical technique that enhances Raman scattering of chemical species adsorbed on rough surfaces or nanoscale structures. It provides s a structural fingerprint by which molecules can be identified. The Raman vibrations of molecules are in general very weak. However, the Raman signature of molecules that are near metal nanoscale surfaces is greatly enhanced in intensity compared to the molecules alone [1, 5, 6]. The most commonly metal nanoscale surfaces are silver and gold which allow chemical detection of analytes at very low concentration with the potential for single-molecule detection. The enhancement effects observed in the SERS are the results of two different effects: (a) the electromagnetic (EM) effect, enhancement of the local electromagnetic field incident on an adsorbed molecule at a metallic surface, (b) and the chemical effect, due to the electronic resonance/charge transfer between a molecule and a metal surface, which leads to an increase in the polarizability of the molecule [1, 5, 6].

More recently, silver [1, 5, 6] and rhodium nanoparticles [7] (NPs) have been investigated for sensing via surface-enhanced Raman (SERS) scattering technique. Bulk Rh–Ag alloys are thermodynamically unstable and therefore do not mix. At the nano level, however, bottom-up manufacturing makes it possible to mix immiscible metals such as Rh and Ag. Nanoparticles also have higher surface-to-volume ratios compared to bulk metals, leading to an increase in reactivity. Rh–Ag nanoparticle alloys were synthesized via the co-reduction of the metal precursors and the ratio of the two metals was altered to vary the final Rh–Ag composition. Rh–Ag NPs also have a plasmon band that stretches into the UV region. As a result, these NPs can also be used for UV plasmonic applications, including the destruction of toxic molecules or label-free detection of biomolecules [7].

In this paper, we show that silver–rhodium bimetallic nanoalloy particles can be prepared by a simple polyol approach via the co-reduction of the metal precursors at elevated temperature. These hybrid nanomaterials were evaluated for their ability to serve as substrates for surface-enhanced Raman scattering.

Experimental

Materials

All materials such as polyvinyl pyrrolidone (PVP, MW = 40 K), ethylene glycol (EG), rhodium (II) acetate, rhodium (III) acetate, silver nitrate, 4-mercaptophenol (4 mPh), acetone were purchased from Sigma–Aldrich and used as received. All glassware was cleaned with aqua regia and thoroughly rinsed with deionized ultra-filtered water.

Nanoparticle Synthesis

Rh–Ag NPs

In a 50 mL Erlenmeyer flask, 111 mg polyvinylpyrrolidone (PVP) and 10 mL ethylene glycol (EG) were mixed and heated to 150–170 °C. In a separate vial, 13 mg Rh(III) acetate and 19 mg $AgNO_3$ (for 50/50 Rh/Ag) were dissolved in 1 mL water. The aqueous Rh/Ag mixture was swiftly added to the PVP/EG mixture at 150–170 °C. Once the solution turned black, indicating the formation of Rh–Ag particles, the solution was removed from the heat and cooled to room temperature. 5 mL Rh–Ag and 10 mL acetone were added to a falcon tube and centrifuged at 6000 rpm for 30 min. The supernatant was removed, and the particles were resuspended in water. The purification procedure was repeated 1–2 more times, and the Rh–Ag particles were re-suspended in water.

RhNPs

7–20 mg Rh (III) acetate was dissolved in water. In a separate 50 mL Erlenmeyer flask, 111 mg PVP and 10 mL ethylene glycol were heated to 140 °C before the Rh (III) acetate solution was added. The solution was cooled to room temperature and cleaned using acetone. The final RhNPs were re-suspended in water.

AgNPs

111 mg PVP was dissolved in 10 mL ethylene glycol and heated to 170 °C, after which 25 mg $AgNO_3$ in 1 mL water was quickly added. The solution was removed from the heat after it turned yellow. AgNPs were purified with acetone 2 × at 6000 rpm for 60 min and resuspended in water.

Nanomaterials Characterization

The morphologies of the nanoparticles were monitored by scanning electron microscopy (SEM, Hitachi SU8200). The chemical compositions were analyzed using energy-dispersive X-ray analysis (EDS, Oxford Instruments). The optical properties of the nanoparticle solutions were measured using UV–Visible–near-infrared spectroscopy (UV–Vis—NIR, Tec5 MultiSpec).

RAMAN Studies

Surface-enhanced Raman spectra were collected using a laser with an adjustable output and a wavelength $\lambda = 532$ nm (Del Mar Photonics, DMPV-532−1, beam diameter focused to ~20 μm), where the beam path is directed onto the top surface of the solution contained in the methacrylate cuvette. The SERS analyte, 4-mercaptophenol, was added to the nanoparticle solutions and allowed several hours for incubation. Different concentrations of 4MPh were investigated: 0.05, 5, 50, and 1000 uM.

Results and Discussion

Synthesis and Characterization of Rh–Ag

RhNPs, AgNPs and Rh–Ag alloy NPs were synthesized using the polyol synthesis [8, 9]. Specifically, nanomaterials were prepared via the co-reduction of the metal precursors by polyol at elevated temperature. In this reaction, the polyol serves as both: reducing and capping reagent. Polyols have high boiling points which allow the synthesis of nanoparticles at high temperatures without the need of using other extraneous experimental parameters, such as high pressure. The metal precursors are often readily soluble in the polyol which makes this a simple and versatile procedure for mono- and bimetallic nanomaterial preparation. Careful selection of the ratio between reactants introduced in the reaction pot and reaction's temperature can be used for tailoring nanomaterial's final size and morphology. It is expected that higher temperature would make this reaction more thermodynamically favorable. As shown in Fig. 1, aqueous metal precursor solutions are added to a hot solution of poly(vinylpyrrolidone) (PVP) and ethylene glycol (EG), which leads to the formation of nanoparticles via the reduction of the metals by PVP and EG. In the case of Rh–Ag NPs, the co-reduction of the two metals leads to creation of bimetallic nanoparticles. This synthesis was also performed with just Rh or Ag precursors to create RhNPs and AgNPs, respectively.

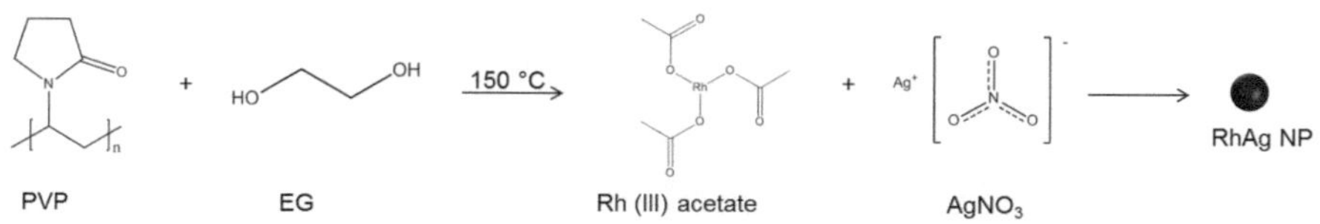

Fig. 1 Polyol synthesis of Rh–Ag alloy NPs

The Rh–Ag NPs display a plasmon band around 410 nm, as shown in Fig. 2. The UV–Vis spectra of the bimetallic nanostructures differ not just from the monometallic nanomaterials but are also highly dependent on the bimetallic composition. If bimetallic Ag–Rh NPs are formed instead of monometallic NPs, then the usual expectation is that the plasmon band will be intermediate between those for the pure metals [5, 6]. Rh NPs do not display plasmon bands in the Vis region of the spectrum. The plasmon band of silver nanospheres is at around 420 nm (data not shown) [1, 5, 6]. This blueshift of the plasmon band suggests that nanomaterials formed are bimetallic and not individual monometallic colloids. Scanning electron microscopy shows that AgRh nanoparticles are spherical with dimension of 21 ± 4 nm in diameter. Figure 2b shows a fairly uniform distribution in size.

The bimetallic nanomaterial composition can be tailored through the precise control of the starting reactants by simply changing the molar ratio of [Rh]/[Ag] precursor solutions. As shown in Fig. 3, the final Rh/Ag weight percent ratio was measured through EDS. For this sample, the nominal Rh–Ag weight percent added in the synthesis was 25% Rh–75% Ag, but the EDS recorded 8% Rh–92% Ag, indicating that the $AgNO_3$ is more easily reduced compared to the Rh (III) acetate. The EDS mapping data also shows that the final product is a hybrid nanomaterial and not a mixture of individual Au and Rh NPs. This agrees with the UV–Vis data reported above.

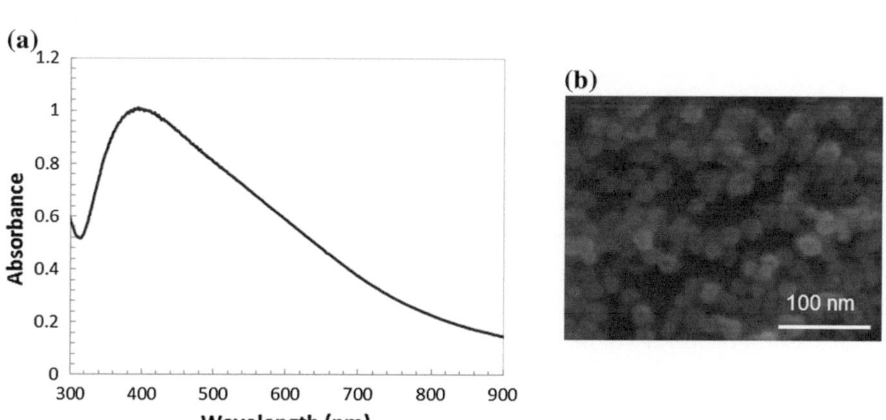

Fig. 2 **a** UV–Vis and **b** SEM of Rh–Ag alloy NPs ($d_{\text{Rh–Ag}} = 21 \pm 4$ nm)

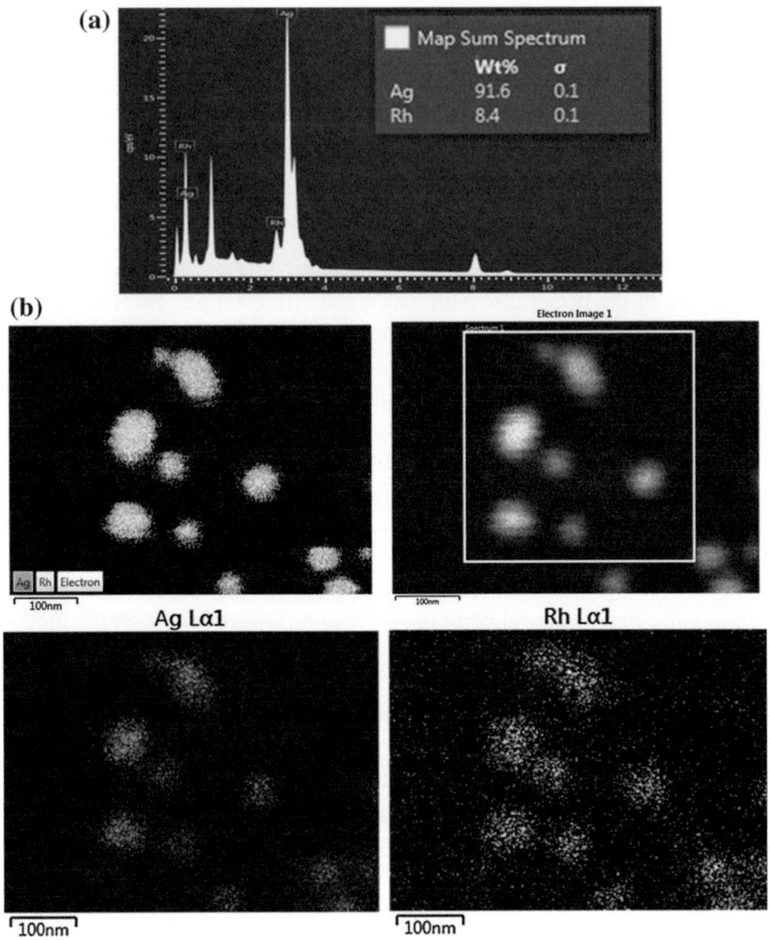

Fig. 3 (top) EDS and (bottom) EDS mapping of Rh–Ag alloy NPs; **a** AgRh EDS map, **b** Rh EDS map and **c** Ag EDS map

Surface-Enhanced Raman Spectroscopy with Rh–Ag Alloys

Here, we used 4-mercaptophenol (4-mPh) as a model analyte for surface-enhanced Raman spectroscopy (SERS) studies. 4-mPh molecules are attached on the metal surface as a monolayer via metal–S bond. As shown in Fig. 4, all the spectra show the characteristic peaks of the 4-mPh spectrum. The peak at 1080 cm^{-1} is the ring breathing mode n1 (Wilson notation), and the other peaks at 391, 635, 824, 1009, 1173, 1494 and 1596 cm^{-1} correspond to the ring vibration mode of 7a, 12a, 6a, 18a, 9a, 19a, and 8a of MPh, respectively [5, 6].

For a concentration of 1.4 mM 4-mercaptophenol (4-mPh), no Raman signal was recorded. However, when the same concentration of 4-mPh was used while attached

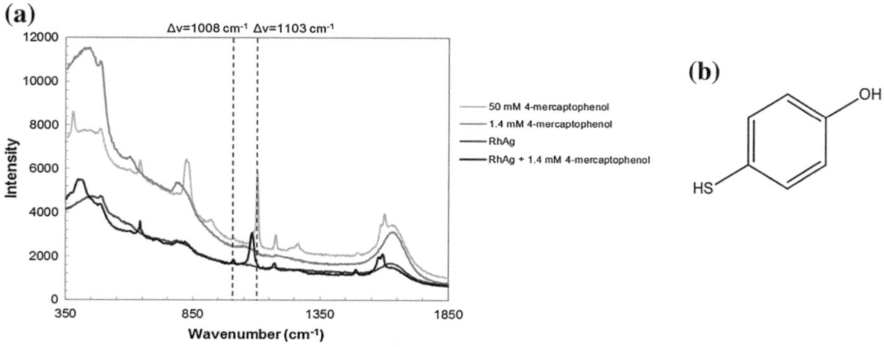

Fig. 4 **a** Raman of Rh–Ag NPs with/without 4-mPh; **b** structure of 4-mercaptophenol

on the Rh-Ag bimetallic nanostructures, a Raman enhanced response was recorded. The Raman peaks of the SERS-enhanced signal were also slightly blueshifted compared to the native solution. This is probably due to the presence of defects and lattice mismatch between Ag and Rh [10].

In Fig. 4, the Raman frequency for the in-plane ring vibrational mode at 1103 cm^{-1} for 4-mercaptophenol is also blueshifted ~20 cm^{-1} when attached to Rh–Ag NPs. This has been reported by others and is attributed to charge transfer or local bonding effects that affect the energies of the vibrational and electronic modes of Raman [10, 11]. For example, gold nanoparticles (AuNPs) have been found to transfer charge with aromatic thiol ligands, such as phenylthiol, where the covalent bond between the ligand and the AuNP leads to an increase in the surface electrons that are responsible for the plasmon band [12]. This is one possible explanation for the observed Raman frequency shift. The peak at 1008 cm^{-1} doesn't shift but becomes more intense in the SERS sample. The analytical SERS enhancement factor (EF) was calculated based on Eq. 1 [1, 5, 6, 13].

$$\mathrm{EF} = \frac{I_{\mathrm{SERS}}/c_{\mathrm{SERS}}}{I_R/c_R} \tag{1}$$

where I_{SERS} and c_{SERS} are the intensity and concentration of the sample with the NPs, and I_R and c_r are the intensity and concentrations in the absence of the NPs, respectively. For the Rh–Ag NPs in Fig. 4, the EF was calculated as 19.

The effect of Rh–Ag NPs concentration on the SERS was investigated. As expected, as the concentration of Rh–Ag NPs increased, the intensity for the ring vibration of 4-mercaptophenol also enhanced. A linear enhancement response was demonstrated (data not shown). This suggests that a monolayer of 4-mPh is achieved and the solutions are sufficiently diluted and does not generate uncontrolled Raman enhancement due to molecule resigning in nanoparticles junctions and aggregates [5].

The SERS response of the 4-mPh concentration from 0.05 to 1000 uM was further examined. As shown in Fig. 5, the lower 4-mPh concentrations have a greater Raman

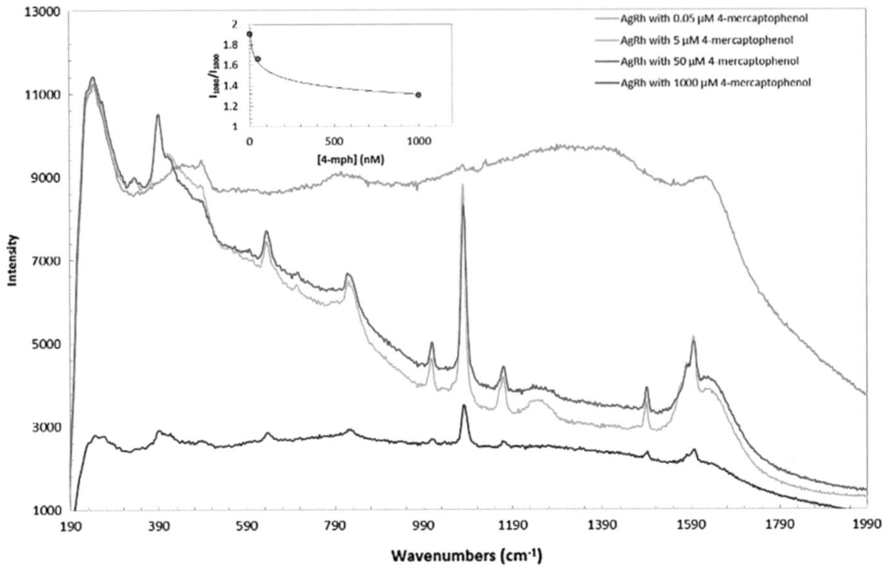

Fig. 5 SERS for different [4 mPh]; Inset: scatter plot of ratio of I(1080 cm^{-1})/I(1008 cm^{-1}) as a function of [4 mPh]

enhancement, but below 5 µM, a Raman signal was not observed. This suggests that a concentration of 5–50 uM leads a monolayer formation on the nanoparticle's surface and is suffice for a recordable SERS response.

For comparison, the SERS response of the control nanoparticles, AgNPs and RhNPs was recorded. As shown in Fig. 6, SERS enhancements were recorded when AgNPs were used. The AgNPs themselves provide a five-order magnitude

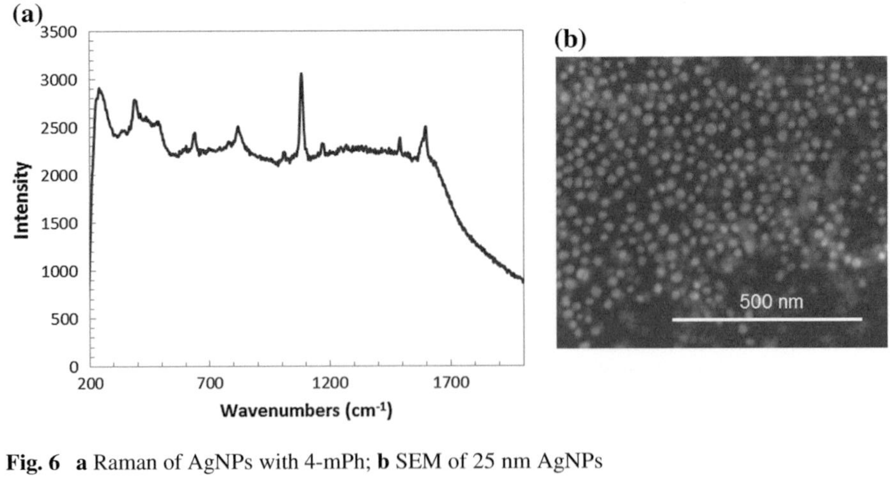

Fig. 6 **a** Raman of AgNPs with 4-mPh; **b** SEM of 25 nm AgNPs

enhancement. Our group has previously reported surface-enhanced Raman scattering of 10^7 compared to silver nanowires when evaluated against the analyte alone, which depended on the composition of the substrate [1, 5, 6]. This is not surprising as nanoparticle's shape, such as rods, triangles, etc., can improve the enhancement factor of SERS by two orders of magnitude compared to spheres. No response was recorded with the RhNPs alone. In our study, the enhancement factors of 5–19 orders of magnitude compared to the Raman spectrum of pure solid 4-MPh were recorded for the bimetallic nanostructures. The EF for the AgNPs was 5, whereas for Rh–Ag it was 19, demonstrating that the bimetallic nanostructures are better SERS-responsive nanomaterials. This is due to the possible formation of pinholes in the lattice mismatch which could function as SERS "hotspots" leading to enhanced response.

Conclusions

Rh–Ag bimetallic NPs were successfully prepared via the co-reduction of the metal precursors by polyol at elevated temperature and characterized via SEM, UV–Vis, EDS and EDS mapping. The Rh–Ag bimetallic nanoparticles demonstrated successful SERS scattering effects, with an enhancement factor almost four times higher than for AgNPs. This suggests that, for measuring low concentrations of analyte by SERS, AgRh NPs provide an additional advantage compared to AgNPs or RhNps.

Acknowledgements This work was supported by the Laboratory Directed Research and Development (LDRD) program within the Savannah River National Laboratory (SRNL). Work was conducted at SRNL under the U.S. Department of Energy Contract DE-AC09-96SR18500.

References

1. Hunyadi Murph SE, Larsen G, Coopersmith K (2017) Anisotropic and shape-selective nanomaterials: structure-property relationships, nanostructure science and technology series. Springer Publisher, pp 1–470
2. Hunyadi Murph SE, Heroux K, Turick C, Thomas D (2012) Metallic and hybrid nanostructures: fundamentals and applications. In: Applications of nanomaterials, Vol 4, ISBN: 1-62699-000-X. In: Govil JN (ed) Nanomaterials and nanostructures, Vol 4, ISBN: 1-62699-004-2, Studium Press LLC, USA
3. Hunyadi Murph SE, Larsen GK, Korinko P, Coopersmith KJ, Summer AJ, Lewis R (2017) Nanoparticle treated stainless steel filters for metal vapor sequestration. JOM 69:162–172
4. Larsen G, Farr W, Hunyadi Murph SE (2016) Multifunctional Fe_2O_3–Au nanoparticles with different shapes: enhanced catalysis, photothermal effects, and magnetic recyclability. J Phys Chem C 120:15162–15172
5. Hunyadi SE, Murphy CJ (2006) Bimetallic silver-gold Nanowires: fabrication and use in surface- enhanced raman scattering. J Mater Chem; Special Issue: Anisotrop Nanoparticles 16:3929–3935
6. Hunyadi Murph SE, Murphy CJ (2013) Patchy Silica-Coated Silver Nanowires as SERS substrates. J Nanoparticle Res 15(6):1607

7. Zhang X, Li P, Barreda A, Gutiérrez Y, González F, Moreno F, Everitt HO, Liu J (2016) Size-tunable rhodium nanostructures for wavelength-tunable ultraviolet plasmonics. Nanoscale Horiz 1:75–80
8. Sun Y, Xia Y (2002) Shape-controlled synthesis of gold and silver nanoparticles. Science 298:2176–2179
9. Biacchi AJ, Schaak RE (2011) The solvent matters: kinetic versus thermodynamic shape control in the polyol synthesis of rhodium nanoparticles. ACS Nano 5:8089–8099
10. Kao P, Malvadkar NA, Cetinkaya M, Wang H, Allara DL, Demirel MC (2008) Surface-enhanced raman detection on metalized Nanostructured Poly(p-xylylene) films. Adv Mater 20:3562–3565
11. Chase SJ, Bacsa WS, Mitch MG, Pilione LJ, Lannin JS (1992) Surface-enhanced Raman Scattering and Photoemission of C60 on noble-metal surfaces. Phys Rev B 46:7873–7877
12. Goldmann C, Lazzari R, Paquez X, Boissière C, Ribot F, Sanchez C, Chanéac C, Portehault D (2015) Charge transfer at hybrid interfaces: plasmonics of Aromatic Thiol Capped Gold Nanoparticles. ACS Nano 9:7572–7582
13. Le Ru EC, Blackie E, Meyer M, Etchegoin PG (2007) Surface enhanced raman scattering enhancement factors: a comprehensive study. J Phys Chem C 111:13794–13803
14. Song C, Abell J, He Y, Hunyadi Murph SE, Cui Y, Zhao Y (2012) Gold-Modified silver nanorod arrays: growth dynamics and improved SERS properties. J Mater Chem 22:1150–1159
15. Watson AM, Zhang X, Alcaraz de la Osa R, Sanz JM, González F, Moreno F, Finkelstein G, Liu J, Everitt HO (2015) Rhodium nanoparticles for ultraviolet plasmonics. Nano Lett 15:1095–1100

New Electron-Beam Equipment and Technologies for the Production of Advanced Materials Using Vacuum Melting and Evaporation Methods Developed at SPE [“Eletekhmash”]

N. I. Grechanyuk, P. P. Kucherenko, A. G. Melnik, I. N. Grechanyuk, Yu. A. Smashnyuk, V. G. Grechanyuk and A. Manulyk

Abstract We present the results of the electron-beam equipment and technologies developed to produce materials and coatings, performed in Scientific and Production Enterprise “Eltekhmash” (Ukraine) in the period from 2005 to 2017. The presentation is about result analysis of developments of a new generation of electron-beam equipment and technologies in this company over more than 10 years. The company is intensively developing such directions of electron-beam technology:

- Production of electric contacts;
- Gas turbine blades protective coating manufacturing;
- Development of a variety of laboratory and production electron-beam equipment with different functional capabilities, which are currently realized in L-2 unit, allows saving time and money for the development of a new technological process.

This unit is capable of realizing four types of different technological processes:

- Coating depositions of numerous ingredients on the turbine blades;
- Production of composition materials from the vapor phases including porous and microlaminate types of materials;
- Production dispersed metals, ceramic, and composite powders.

The company is also manufacturing different types of equipment based on mentioned development or equipment according to customer design and goals.

Keywords Electron-beam deposition · Vapor phase · Composites · Gas discharge e-guns with cold cathode

N. I. Grechanyuk (✉) · P. P. Kucherenko · A. G. Melnik · I. N. Grechanyuk · Yu. A. Smashnyuk · V. G. Grechanyuk
Kiev, Ukraine
e-mail: eltechnic777@ukr.net

A. Manulyk
Oshawa, Canada

T. S. Srivatsan and M. Gupta (eds.), *Nanocomposites VI: Nanoscience and Nanotechnology in Advanced Composites*, The Minerals, Metals & Materials Series, https://doi.org/10.1007/978-3-030-35790-0_9

Introduction

Advanced material design needs a variety of properties, lower cost, and non-limitation for development. The building of the necessary compositions from elements on micro- and nanolevels, like atoms/molecular condensation in a vacuum on the substrates more and more, occupies material development. The simplicity of the creation of functional parts for airspace, automotive, medicine in the form of coatings or bulk constructed parts through evaporation of components is evident. Another advantage of the proposed methods is a combination polarity of properties in the bulk or coating products.

HEEBPVD quick evaporation by electron beam guns discharge e-guns (HEEBPVD) along with a plasma sintering process of new compositions in vapor clouds is part of the philosophy and allows for the production of a wide specter of the new materials for different industries. The goal of newly created production/research enterprise is a production of a variety of materials in the form of a coating or thin bulk plates of different compositions and sizes, which will be used in a variety of industrial applications.

Growing material demands create more opportunities for future companies. The proposed product of new thick coating with a variety of properties such as high corrosion, erosion, and oxidation resistance along with an ability to get same compositions in a bulk form satisfies customer demands with same product prices with higher qualities and high production rates.

Main Milestones of SPE "Eltekhmasch"

In this work, we present results of the development of electron-beam equipment and technologies for producing materials and coatings made in scientific-production enterprise "ELETEKHMASH" in the last decade. The company is intensively developing such direction of electron-beam technology:

- Design and mounting laboratory and industrial equipment for melting of metals and alloys, production of protective coatings, production of composite materials condensed from a vacuum phase;
- Production of high purity Ni–W alloys used as seeds in growing single-crystal blades;
- Production of titanium alloys for biomedical applications;
- Master alloy production;
- Production of high-quality ingots from a scrap of high-temperature alloys IS26-VI and JS32;
- Manufacturing tubular billets-cathodes from Ni–Cr–Al–Y, Ni–Co–Cr–Al–Y heat-resistant alloys for ion-plasma coating deposition;
- Production of metal powders for thermo-spray and cold-spray depositions;
- Production of electrical contacts;
- Production of protective coatings on gas turbine blades.

Fig. 1 Unit L-2 (From The Paton Welding Journal, No. 5–6, 2016. Used with permission)

Many years' experience opens a possibility to design and build multi-functional electron-beam machine, which can perform a lot of technological processes in one technological chamber and allows to save time and money for the new development of materials and technological processes. Recently, a new semi-production/semi-experimental unit L-2 was proposed to the market, see Fig. 1.

This unit allows the realization of four types of different technological processes. The first of them is the coating deposition on various parts in tribune blades, Fig. 2. Three independent water-cooled crucibles with diameters of 70 mm make evaporations independently (simultaneously) of three different materials according to the programmed process. Such a design with strict control gives an ability to get heat-resistant, metal, ceramic, single-layer, and multilayer graded coatings. There is a two-chamber variant design for better vacuum keeping. Loaded and unloaded parts in the chamber are performed without vacuum breaking due to the separation of the chamber in two chambers: loading/unloading chamber and technological chamber.

Microlaminated, dispersion-strengthened, and microporous types of composite materials produced from the vapor phase allow the unit to perform the second task. At the evaporation from three independent crucibles, the vapor flow is deposited on stationary or rotated substrates with 500 mm in the diameter, Fig. 3. The thickness of the coating can reach up to 20 mm. For easy separation of coated (sintered) material substrate is covered and advanced with thin non-diffusion composition. Standard composition thickness is between 1 and 7 mm (Table 1).

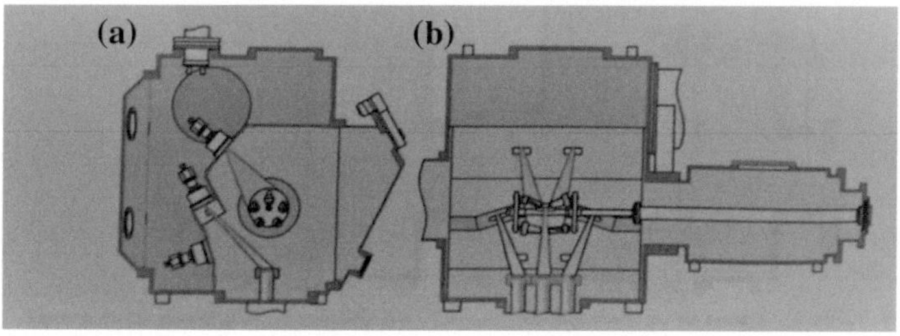

Fig. 2 Schematic of coating deposition on gas turbine blades **a** side view, **b** front view (From The Paton Welding Journal, No. 5–6, 2016. Used with permission)

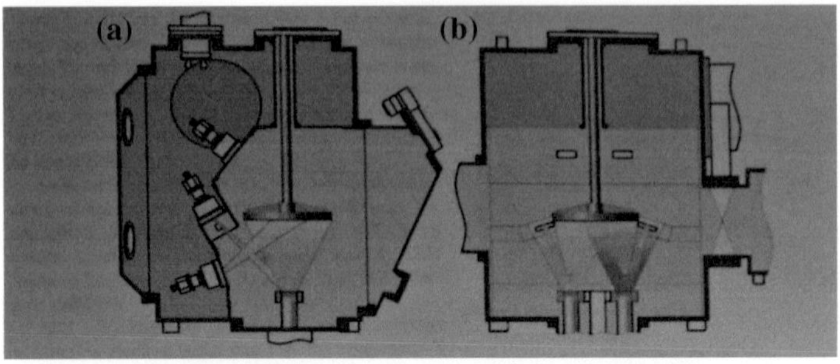

Fig. 3 Schematic of production composite materials condensed from vapor phase **a** top view; **b** front view (From The Paton Welding Journal, No. 5–6, 2016. Used with permission)

The new technological process developed in the L-2 is dispersed metal, ceramic composites, and composite powders.

At present much attention is given to the development of thermo-barrier coatings (TBC) on the blades. SPE Elteckmash designed and built new electron-beam unit L-8 for TBC deposition on turbine blades. Figure 4 and schematic of the technological process in Fig. 5 L-8 unit are loaded with gas discharge e-guns, and operating life of them is up to 1000 h, and they apply for the deposition of any types of protective coatings: metal, ceramics, composites, with a single layer, multilayers, graded layers. All complex compositions of TBC layers can be formed in one processing circles. There are two loading/reloading lock chambers supplied with ion cleaning of blades before coating deposition in these chambers' blades can be preheated as well. Here as well, the formation of barrier microlayers with inner heat-resistant layer and outer ceramic layer for diffusion slow down in interfaces also takes place. For controlling coating thickness, there is a sensor mounted on the technological chamber wall along with sighting tube and viewing window for contactless measurements of the item's temperature. For that purpose, high-temperature infrared pyrometer is mounted on the working chamber rear wall. The unit also supports partial technological gas ionization and metal vapors by applying a negative bias to the parts (up to 2 kV) for better adhesion of the coating and its quality improvement.

Production of Electrical Contacts

The company commercialized new material, Cu–Zr–Y–W and Cu–Zr–Y–Cr, for electrical contacts:

This material provides high reliability of contacts operation, and the maximum value of the switched current is up to 1200 A;

- It is properly processed by cutting, punching, grinding, and drilling, and it is easy to solder for any of well-known ways of the soldering using standard silver-consist and not silver-consist solders;

Table 1 Specification of EB unit L-2 (From The Paton Welding Journal No. 5–6, 2016. Used with permission)

Specification name	Parameter value
Dimensions of evaporated billets melted from the upper mechanism, mm	
Diameter Length	80 390
Detentions of condensed surfaces, mm	
Rectangular Round	350 × 350 400
Cylindrical	
Diameter Length	200 350
Distance from evaporation surface to condensation surface, mm	200–325
Number of crucibles, pcs	3
Speed of evaporation, ingot displacement, mm/min	1–350
Speed of displacement of billets melted from the top, mm/min	5–350
Speed of substrate holder rotation on a horizontal shaft, rpm	3–25
Speed of substrate holder rotation on a vertical shaft, rpm	5–70
Number and power (kW) of electron-beam guns (thermal cathode guns with strip cathode):	
– for material evaporation from the crucibles – for heating from above – for heating from below	3 × 60 2 × 60 1 × 60
Consumed power, kW, not more than	
High voltage power source The power source of an ion cleaning device Rated accelerated voltage, kV	250 30 20
Working vacuum in the chamber, Pa (mm Hg)	6×10^{-3} 1×10^{-2} 5×10^{-3} 1×10^{-4}
Dimensions of the unit, mm, not more than	
Length Width Height Unit weight, tons	4300 6200 3300 16.5

- Contacts have had high radiation stability, thermal stability, friction resistance, have high heat and electrical conductivity;
- Material is not supporting arc burning;
- Punching, grinding, drilling of the material are easy as well to solder by using standard silver-consist and not silver-consist solders.

A company manufactured around 1.6 million electrical contacts of different sizes for national and international economy applications.

Fig. 4 TBC electron-beam unit L-8 (From The Paton Welding Journal, No. 5–6, 2016. Used with permission)

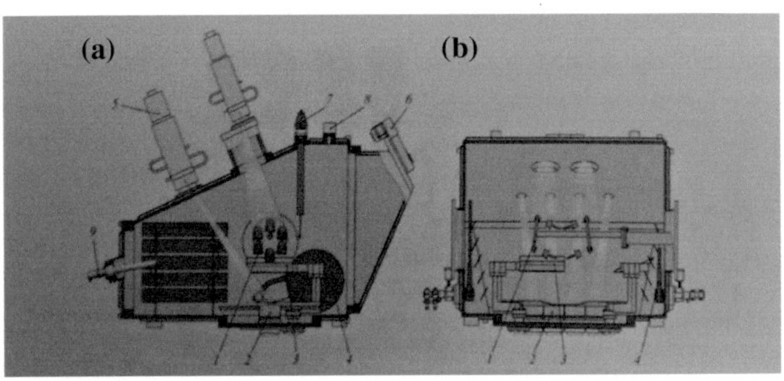

Fig. 5 Schematic of TBC technological process in unit L-8 **a** cross-section view, **b** longitudinal view, 1-holder with blades, 2-crucible, 3-evaporator gate valves, 4-lock gates valves, 5-electron-beam gun, 6-viewing system, 7-load cell, 8-process gas leak valve, 9-ball lead for pyrometer mounting (From The Paton Welding Journal, No. 5–6, 2016. Used with permission)

Coating Depositions of Gas Turbine Blades

Scientific Production Enterprise "ELTECHMASH" is manufacturing coatings on turbines blades obtained by electron-beam evaporation of alloys like MeCrAlY (where Me is Ni, Co, Fe), MeCrAlYHfSiZr and ceramic on the ZrO_2 base, stabilized by Y_2O_3 with further steam phase condensation on the surface of gas turbine blades.

Three classes of coatings are developed:

- One-layer metal coatings such as MeCrAlY and MeCrAlHfSiZr;
- One-layer composite types such as with alternation of layers MeCrAlY/(MeCrAlYHfSiZr) MeCrAlY/(MeCrAlYHfSiZr)+MeO, (where MeO is Al_2O_3 or $ZrO_2 + 6–8\%\ Y_2O_3$);

Table 2 Specification of EB unit L-8 (From The Paton Welding Journal No. 5–6, 2016. Used with permission)

Dimensions of the cylindrical cassette with parts, mm, not more than	
Diameter Length	250 500
Speed of item rotation on a horizontal shaft, rpm	05–50
The number of evaporators, ps.	4
Crucible inner diameter, mm	70
Length of evaporators ingots, mm, not more than	500
Ingot feed rate, mm/min	05–350
Distance from the upper edge to cassette rotation axis or condensation plane, mm	350
Number of nominal power (kW) of electron-beam guns	
For material evaporation from the crucible For item heating	4×100 2×60
Type of electron-beam guns—axial guns with the cold cathode (based on high voltage gloving discharge)	
Consumed power, kW, not more than	
High voltage power sources Auxiliary equipment	520 80
Rated accelerated voltage, kV	30
Working vacuum in the chamber, Pa (mm Hg)	6×10^{-3} $6 \times 6 \times 10^{-2}$ 5×10^{-3} 5×10^{-4}
Unit overall dimensions, mm, not more than	
Length Width Height	10,500 9500 4300
Weight of the unit (with power source) not more than tone	25

- Two-layer coatings with internal metal MeCrAlY/(MeCrAlYHfSiZr) and external ceramic (ZrO_2–Y_2O_3);
- Two-layer coatings with internal composite MeCrAlY/(MeCrAlYHfSiZr)+MeO dispersion-hardening or microlayer types and external ceramic (ZrO_2–Y_2O_3) layers;
- Three-layer coating with internal and intermediate metal layers on the MeCrAlY/(MeCrAlYHfSiZr) and external ceramic (ZrO_2–Y_2O_3);
- Three-layer coating with internal metal MeCrAlY, (MeCrAlYHfSiZr) intermediate composite MeCrAlY/(MeCrAlYHfSiZr)/MeO.

Company Eltekhmash successfully established a close cycle of coating deposition on gas turbine blades, which included melting and purification of all types of Ni- and Co-based alloys [1] in the form of ingots, manufacture of ZrO_2–Y_2O_3 ceramic billets, following the deposition of obtained materials on gas turbine blades. The company

Table 3 Electron-beam equipment supplied to the customers from 2005 through 2014 (From The Paton Welding Journal No. 5–6, 2016. Used with permission)

Unit name	Application	Year	Customer country
Electron-beam unit L-1	Deposition of protective coatings from the vapor phase in vacuum	2005	Ukraine
Electron-beam unit L4	Refining and melting of metals and alloys in vacuum	2006	Armenia
Two power units with HVGD-based guns of 220 kW each	Commercial production of "solar silicon" from metallurgical silicon	2007	Japan
Power unit with HVGD-based guns of 30 kW and 100 kW of power	Upgrading of the customer machines for metallurgical silicon refining	2008	Taiwan
Power unit with HVGD-based gun of 100 kW of power	Unit upgrading for refining and melting noble metal alloys	2010	Russia
Electron-beam unit L2	Deposition of protective coating from the vapor phase in vacuum	2012–2013	Ukraine, China
Electron-beam unit L8	Deposition of protective coatings on gas turbine parts	2014	Russia

can produce deposition materials for the customers manufacturing thermo-barrier coating (TBC) on their equipment [1–4] (Table 2).

Manufacturing of Industrial Electron-Beam Equipment

In Table 3, it is possible to see electron-beam equipment supplied to the different countries in the period of 2005–2014. The enterprise supplies production and experimental types of equipment as well as separate parts and block according to the customer demands or design. Complete set of the machines allows realizing technological processes of coating, composition materials production from the vapor phase, and manufacturing of casting high-quality alloys

Conclusions

1. SPE Eltehmash is expanding its business and looking for partners and customers around the world.
2. We are ready for your challenges and interests in new material development and manufacturing.

3. Our specialists are working in different countries, promoting our products and technologies.
4. Our common product can be the next generation of material for growing customer demands.

References

1. TU 27.4-201134.10.0002-2001 (2015) Materials in ingots and powders for protective coatings. Modification N3 to KTU Version 3 of 03.09.2015
2. Grechanyuk NI, Kucherenko PP, Grechanyuk IN (2007) New Electron beam equipment and technologies of producing advanced materials and coatings. Paton Weld J 5:25–29
3. Grechanyuk NI, Kucherenko PP, Melnik AG (2014) Industrial electron beam installation L-8 for deposition of high protective coatings on gas turbine blades, vol 10, pp 45–50
4. Grechanyuk NI, Kucherenko PP, Melnik AG (2016) Industrial electron beam installation L-4 for vacuum re-melting and refining of metals and alloys. Proshkovaia Metall 7(8):140–149

Role of Rare Earth Oxide Reinforcements in Enhancing the Mechanical, Damping and Ignition Resistance of Magnesium

Milli Suchita Kujur, Vyasaraj Manakari, Gururaj Parande, Mrityunjay Doddamani, Ashis Mallick and Manoj Gupta

Abstract Magnesium based nanocomposites, on account of their excellent dimensional stability coupled with mechanical integrity, have provided the much-needed impetus for utilization in both aerospace-related and automobile-related applications. However, the perceived easy ignition and flammability of magnesium alloys create a detrimental safety feature that hinders the aerospace application opportunities. Incorporation of rare earth metal oxides into magnesium matrix can induce 'reactive element effect' (REE), due to their strong rare earth–oxygen interactions. Along with enhancing the protective characteristics of oxides on many metals and alloys, the addition of such rare earth oxides also helps in realizing a refined microstructure and good strength–ductility combination in the composites. This manuscript presents the mechanical properties, damping and ignition resistance characteristics of the new and improved composite materials engineered by reinforcing magnesium with rare earth oxide nanoparticle. Rationale for the observed properties is discussed while concurrently establishing the relationship between microstructure of the engineered composites and resultant mechanical properties.

Keywords Magnesium · Nanocomposites · Rare earth oxide · Microwave sintering · Mechanical properties · Damping behaviour · Ignition resistance

---

M. S. Kujur (✉) · A. Mallick
Department of Mechanical Engineering, Indian Institute of Technology (Indian School of Mines), Dhanbad, Jharkhand, India
e-mail: millisuchitakujur@gmail.com

M. S. Kujur · V. Manakari · G. Parande · M. Gupta (✉)
Department of Mechanical Engineering, National University of Singapore, 9 Engineering Drive 1, Singapore 117576, Singapore
e-mail: mpepgm@nus.edu.sg

M. Doddamani
Advanced Manufacturing Laboratory, Department of Mechanical Engineering, National Institute of Technology Karnataka, Surathkal, Karnataka, India

T. S. Srivatsan and M. Gupta (eds.), *Nanocomposites VI: Nanoscience and Nanotechnology in Advanced Composites*, The Minerals, Metals & Materials Series, https://doi.org/10.1007/978-3-030-35790-0_10

Introduction

The alarming rate of depletion of crude oil coupled with the increasing norms placed on emissions has in more recent years provided the necessary impetus for lightweight materials for selection and use in a spectrum of applications in the industries of automobile and aerospace. Being the sixth most abundant element in the earth's crust, magnesium with its low density and high specific strength [σ/ρ] promises to be befitting for selection and use in the high-performance industries. However, magnesium does possess a few serious limitations, such as (i) low elastic modulus and (ii) low ductility. In the aviation industry, the low auto-ignition temperature of magnesium is an impediment to its selection for use in key and potential applications [1]. Owing to the recent lift of ban on using magnesium in aircraft interiors, usage of magnesium in an aircraft may significantly increase provided the ignition temperature can be increased [2]. Currently, around 43 pct. of the weight of a seat in an aircraft is comprised of aluminum alloys. Replacing the alloys of aluminum with magnesium offers a reduction in weight as high as 30 pct. In an attempt to overcome these limitations coupled with the need to further enhance mechanical properties, magnesium has been incorporated with nanoparticles, thereby engineering a composite material. This has proven to simultaneously increase both the strength and ductility of magnesium [1].

Nanocomposites can be tailor-made to suit different applications by an appropriate selection of the magnesium alloy, the reinforcement, and fabrication technique used. Such inclusion of alloying elements and nanoparticles in pure Mg has improved the surface reactivity along with the mechanical properties without adversely affecting its ductility by grain size strengthening and solid solution strengthening mechanism [3]. The drawback with the alloying elements is that they impose extra weight, thereby increasing the overall density of the material and certain elements such as rare earth elements and aluminium induce a certain level of toxicity in pure Mg [4, 5]. This limitation can be overcome by incorporating nanoparticles at lower amounts in the base matrix. Studies show that metallic ceramic oxides like Al_2O_3, Y_2O_3, TiO_2, ZrO_2, ZnO, SiO_2 when reinforced in Mg matrix display significant improvement in the strength properties as well as ductility [6]. Incorporation of rare earth metal oxides (REOs) including those of the lanthanum series as well as yttrium and scandium into metallic matrices can induce 'reactive element effect' (REE), due to their strong rare earth–oxygen interactions [7]. Along with enhancing the protective characteristics of oxides on many metals and alloys, the addition of such REOs also helps in realizing a refined microstructure and good strength–ductility combination in the composites [7, 8]. Further, REOs are thermally stable reinforcements and hence can be used in extremely high-temperature applications [9]. Compared to other metallic oxide reinforcements, REOs offer unique advantages in enhancing the performance of Mg composites which may not be realized otherwise.

The current work focuses on the synthesis of Mg–Sm_2O_3 and Mg–CeO_2 nanocomposites by powder metallurgy technique followed by hybrid microwave sintering. The environmentally friendly approach of hybrid microwave sintering helps in achieving better end-application properties with massive reduction in processing time and costs. Further, it does not require additional inert protection gas for sintering. Accordingly, this paper presents the effect of presence of REOs on microstructural, mechanical, damping and ignition properties of pure Mg. Also, the structure–property correlations have been drawn and analyzed.

Experimental Procedure

Materials and Processing

The size of Mg powder procured from Merck, (Darmstadt, Germany) was 60–300 μm with 98.5% purity. Size of Sm_2O_3 and CeO_2 nanoparticles procured from US Research Nanomaterials, Inc., USA was 20–30 nm for both, respectively.

Details specific to the preparation process of the Mg–REO nanocomposites have been reported in our earlier study [10, 11]. To produce the Mg–REO nanocomposites, 1.5 vol.% of Sm_2O_3 and CeO_2 nanoparticles and pure magnesium were precision weighed in an electronic balance and subsequently blended at 200 rpm for full 2 h (120 min) using a planetary ball mill [*Model*: Retsch PM400]. The blended composite powder was then compacted into cylindrical billets that had a height of 40 mm and a diameter of 35 mm at a pressure of 97 bars (50 tons). The compacted billets were then sintered using microwave sintering at 630 °C under ambient conditions. Billets of the microwave sintered magnesium and magnesium nanocomposite were homogenized at 450 °C for 2 h (120 min) prior to hot extrusion at a temperature of 400 °C at an extrusion ratio of 20.25:1 on a 150-ton hydraulic press resulting in rods that measured 8 mm in diameter. Samples of the as-synthesized composites were prepared for further characterization in accordance with procedures detailed in the ASTM standards.

Materials Characterization

To investigate the average matrix grain size, REOs nanoparticle distribution in the Mg matrix and interfacial integrity between the nanoparticles and the Mg matrix, microstructural characterization studies of both pure magnesium and Mg-1.5 vol.% Sm_2O_3 and Mg-1.5 vol.% CeO_2 nanocomposites were conducted on metallographically polished extruded samples using an Olympus metallographic optical microscope (Tokyo, Japan), a Scion image analysis software (Sacramento, CA, USA) and a Hitachi S-4300 Field Emission Scanning Electron Microscope (FESEM) (Tokyo, Japan). For each polished sample, five micrographs [20 grains each] were carefully analyzed for an accurate estimation of the grain size.

The CTE values of the synthesized materials were analysed using a LINESIS TMA PT 1000LT thermo-mechanical analyser (Tokyo, Japan). The argon flow was maintained at 0.1 litres per min (lpm) while heating rate was set to 5 °C/min. The displacement of the samples was measured using an alumina probe in a temperature range of 50–400 °C.

The ignition temperatures of the synthesized materials were determined using a thermo-gravimetric analyzer (TGA). Samples of dimensions 2 × 2 × 2 mm were heated from 30 to 750 °C at a heating rate of 10 °C min^{-1} in the purified air with a flow rate of 50 mL min^{-1}.

The room temperature compressive properties of both extruded pure Mg sample, Mg-1.5 vol.% Sm_2O_3 and Mg-1.5 vol.% CeO_2, were determined in conformance with ASTM E9-89a using an MTS-810 testing machine and a strain rate of 10^{-4} s^{-1}. The test specimens measured 8-mm diameter and 8-mm in length. For each composition, five specimens were tested to ensure repeatability of the results.

The damping performance of pure Mg and Mg–REO nanocomposites was conducted according to the ASTM standard E1876-09 and using a resonance frequency damping analyzer [Model: RFDA, IMEC, Genk, Belgium]. Each sample used for this specific study measured 60 mm in length and 8 mm in diameter.

Results and Discussion

Microstructure

Visual observation of the microwave sintered nanocomposite billets and the extruded rods clearly revealed an absence of any macroscopic surface defects. This confirms an overall suitability of the processing parameters used during the processes of sintering and extrusion. The intrinsic microstructural features of the developed pure magnesium, Mg-1.5 vol.% Sm_2O_3 and Mg-1.5 vol.% CeO_2 nanocomposites were studied in terms of the presence and distribution of the reinforcing REOs nanoparticles through the microstructure. Figure 1 shows the inclusion of Sm_2O_3 and CeO_2 nanoparticles as reinforcements in pure Mg and are seen to be distributed homogeneously throughout the matrix. Careful selection of blending, high extrusion ratio and sintering parameters was capable enough in breaking down the agglomerates and dispersing them in the matrix, although some amount of agglomeration can be still noticed for Mg-1.5 vol.% Sm_2O_3 nanocomposites.

The tabulated form of grain size is provided in Table 1. The grain size of the hot extruded pure Mg was observed to decrease with the addition of REO nanoparticles. Further, the grains were predominantly equiaxed in shape for both the cases. The Mg-1.5 vol.% Sm_2O_3 and Mg-1.5 vol.% CeO_2 nanocomposites exhibited a grain size of ~12.9 and 10.5 μm, which is ~49.8 and 59.1% lower than that of pure magnesium, respectively. The reason for reduction in the grain size can be attributed to the ability of ultrafine REO nanoparticles to pin the grain boundaries [12].

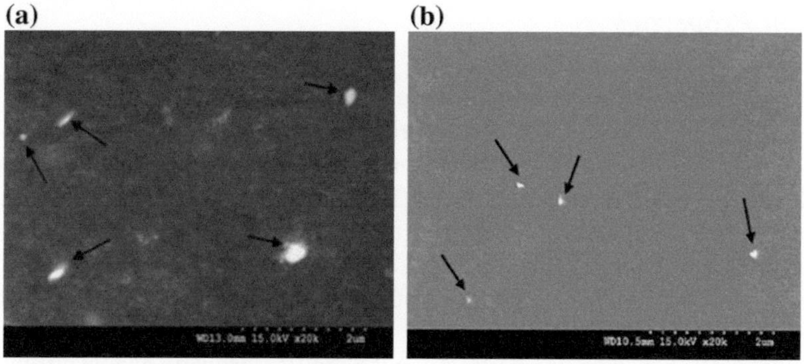

Fig. 1 SEM images for **(a)** Mg-1.5 vol.% Sm_2O_3 and **(b)** Mg-1.5 vol.% CeO_2 showing the distribution of nanoparticles in the Mg matrix taken from [10, 11]

Table 1 Results of grain size measurements

Sample	Grain size (μm)
Pure Mg	25.7 ± 2.6
Mg-1.5 vol.% Sm_2O_3	12.9 ± 2.2
Mg-1.5 vol.% CeO_2	10.5 ± 1.8

Coefficient of Thermal Expansion (CTE)

Results of coefficient of thermal expansion (CTE) measurements made on pure magnesium and nanocomposite samples are provided in Table 2. There was reduction in the coefficient of thermal expansion values for Mg-1.5 vol.% Sm_2O_3 (24.11×10^{-6}/K) and Mg-1.5 vol.% CeO_2 (22.42×10^{-6}/K) nanocomposites in comparison to pure Mg (26.27×10^{-6}/K) as seen in Table 2. The CTE values were reduced by 8.22% and 14% for Mg-1.5 vol.% Sm_2O_3 and Mg-1.5 vol.% CeO_2 when compared to pure Mg. The difference in the thermal expansion coefficient of Sm_2O_3 (10.6×10^{-6}/K) and CeO_2 (8.5×10^{-6}/K) nanoparticles with respect to pure Mg (27.1×10^{-6}/K) led to lower values of thermal expansion values for the nanocomposites. The addition of thermally stable rare earth oxide nanoparticles in pure Mg contributes in maintaining the dimensional stability of the material [10, 11].

Table 2 Results of coefficient of thermal expansion of the synthesized materials

Sample	CTE ($\times 10^{-6}$/K)
Pure Mg	26.27
Mg-1.5 vol.% Sm_2O_3	24.11
Mg-1.5 vol.% CeO_2	22.42

Ignition Properties

According to the FAA, ignition is the point where the white flame appears and starts to propagate [13]. Although the reaction of magnesium with oxygen is exothermic in nature, and releases substantial heat, ignition may not lead to burning if enough heat is removed [13]. The ignition testing here focuses on the temperature at which the material starts to burn. Table 3 reports the ignition temperature for pure Mg, Mg-1.5 vol.% Sm_2O_3 and Mg-1.5 vol.% CeO_2 which is 581 °C, 650 °C and 610 °C, respectively. The thermally stable rare earth oxide, Sm_2O_3 and CeO_2, nanoparticles were able to improve the ignition temperature of pure Mg by 69 °C and 26 °C, respectively [10, 11]. This improvement in the ignition property can be attributed to

(a) the ability of the rare earth oxides to provide a non-conductive interface for heat transfer,
(b) formation of protective oxide layer at high temperatures owing to the addition of rare earth oxides which modify the surface of Mg and
(c) insulation offered by thermally stable oxides on the surface of Mg.

It can be inferred that ignition is governed by oxidative mechanisms [14]. At 440 °C or above, Mg evaporates and reacts with oxygen and forms MgO at the surface of the nanocomposite. This MgO film has pilling–bedworth ratio (PBR) of 0.81 and isn't strong enough to form a protective barrier and would enable formation of oxide nodules [15]. With the addition of Sm_2O_3 and CeO_2 nanoparticles having pilling–bedworth ratio (PBR) of 1.02 and 1.07, respectively, may result in the change of intrinsic chemistry of the base matrix, hence allowing the formation of rich rare earth oxide layer beneath the MgO film in pure Mg [15]. This rich protective rare earth oxide layer would suppress oxide diffusivity characteristics and would prolong the

Table 3 Results of ignition temperature and thermal conductivity for synthesized materials

Material	Ignition temperature (°C)	Thermal conductivity (W/m-k)
Pure Mg	581	135
Mg-1.5 vol.% Sm_2O_3	650	132.97
Mg-1.5 vol.% CeO_2	610	133.151
AM60	525	
AZ61	559	
ZK60A	499	
AZ91	600	
AM50	585	
AZ63	573	
ZK51A	552	
ZK40A	500	

material from igniting early. Also, by incorporating lower thermal expansion coefficient of Sm_2O_3 (10.6×10^{-6}/K) and CeO_2 (8.5×10^{-6}/K) nanoparticles would provide an insulating effect throughout the matrix [10, 11]. The ignition property can also be correlated with thermal conductivity. The thermal conductivity at 400 °C for pure Mg, Mg-1.5 vol.% Sm_2O_3 and Mg-1.5 vol.% CeO_2 is 135 (W/m-k), 132.97 (W/m-k) and 133.151 (W/m-k), respectively, when calculated theoretically by rule of mixtures [10, 11]. Incorporation of REO reinforcements with lower CTE values decreased the CTE values of the nanocomposites as seen in Table 3 in turn offering better thermal and dimensional stability, thus improving the ignition resistance [11]. With the decrease in the thermal conductivity values, there is a rise in ignition temperature as seen in Table 3. It can be suggested that the thermal conductivity of the nanocomposites is directly related to the amount of reinforcement added to the matrix, and in view of the ability of the reinforcement to reduce the availability of the matrix for ignition [10]. Additionally, the fair dispersion of rare earth oxide nanoparticles would delay the ignition by not allowing the heat transfer to take in the matrix across the nanoparticles. The ignition temperature of Mg-1.5 vol.% Sm_2O_3 (650 °C) and Mg-1.5 vol.% CeO_2 (610 °C) nanocomposite is higher than most of the commercially available alloys like AM60 (525 °C), AZ61 (559 °C), ZK60A (499 °C), AZ91 (600 °C), AM50 (585 °C), AZ63 (573 °C), ZK51A (552 °C) and ZK40A (500 °C) [10] and is, hence, suitable for aerospace applications.

Compressive Behavior

The room temperature compression properties of the developed materials are shown in Table 4. The incorporation of Sm_2O_3 and CeO_2 NPs into pure Mg matrix resulted in significant improvement of both 0.2% compressive yield strength (CYS) and ultimate compressive strength (UCS) values. The CYS values of pure Mg improved with the addition of Sm_2O_3 and CeO_2 nanoparticles by ~73% and ~50%, respectively (Table 4). Similarly, the UCS values for Mg-1.5 vol% Sm_2O_3 and Mg-1.5 vol.% CeO_2 nanocomposites enhanced by ~58.63% and ~32.12% compared to pure Mg (Table 4). This significant enhancement in the strengths under compressive loading of nanocomposite can be attributed to synergistic effects of

Table 4 Compressive properties of pure Mg and Mg–REO nanocomposites

Sample	Compressive properties		
	CYS (MPa)	UCS (MPa)	Failure strain (%)
Pure Mg	74 ± 3	249 ± 6	17.4 ± 0.3
Mg-1.5 vol% Sm_2O_3	128 ± 5	395 ± 7	17.2 ± 0.5
Mg-1.5 vol.% CeO_2	111 ± 2	329 ± 7	19.7 ± 0.8

(a) superior refinement of grains by REOs addition (Table 1), promoting the Hall–Petch strengthening mechanism,
(b) uniform dispersion of REO NPs which act as obstacles to dislocation movement through the Orowan strengthening mechanism,
(c) combined effects of texture randomization and deformation twinning creating additional barriers to the crack path leading to enhanced compressive strengths and
(d) effective load transfer from the matrix to the reinforcement, owing to their good interfacial bonding [9].

Further, fracture strain values remained similar to pure Mg in case of Mg-1.5 vol.% Sm_2O_3 nanocomposite, whereas Mg-1.5 vol.% CeO_2 exhibited values of 19.7% which are ~13.3% improved compared to pure Mg. Inclusion of REOs would promote grain alignment, aiding slip as well as deformation twinning. The REO reinforcements assist in hindering the twinning by activation of new slip/twin deformation modes by change in the crystallographic texture in the base matrix, hence improving the fracture strain [10]. Additionally, nanoscale buckling of ceria nanoparticles in Mg matrix aids in dispersing localized stored energy during compressive deformation, hence increasing fracture strain values [11].

Damping Behavior

The vibration damping capabilities of pure Mg and the Mg-1.5 vol.% Sm_2O_3 and Mg-1.5 vol.% CeO_2 nanocomposite are summarized in Table 5. An enhancement in the damping capacity (Q^{-1}) and damping loss rate (L) of the pure magnesium is observed with the addition of Sm_2O_3 and CeO_2 nanoparticles. The Mg-1.5 vol.% CeO_2 nanocomposite revealed the best values of damping capacity and damping loss rate to be ~18.97 × 10^{-4} and 40.9, which is ~381% and 387% enhancement compared to pure Mg, respectively. The damping capacity and damping loss rate for the Mg-1.5 vol.% Sm_2O_3 nanocomposite sample improved by ~189% and ~346% when compared one-on-one with that of pure magnesium (Mg).

Several damping-related mechanisms are responsible for the observed improvement in the damping characteristics of pure magnesium (Mg) with the addition of REO nanoparticles and these include the following:

Table 5 Elastic modulus and damping characteristics of extruded Mg–REO nanocomposites

Sample	Damping capacity (× 10^{-4})	Damping loss rate	Elastic modulus (GPa)
Pure Mg	3.94 ± 0.21	8.2 ± 0.2	42.3 ± 0.14
Mg-1.5 vol.% Sm_2O_3	11.39 ± 0.8	36.65 ± 0.9	44.9 ± 0.2
Mg-1.5 vol.% CeO_2	18.97 ± 1.2	40.9 ± 2.7	43.9 ± 0.02

(a) Bulk texture modification.
(b) Enhanced damping at the interface binding the REO nanoparticles in the magnesium (Mg) matrix, and
(c) High number of dislocations owing to mismatch in the thermal expansion coefficient of the soft and ductile magnesium (Mg) matrix and the reinforcing REO nanoparticles.

The damping response of magnesium-based materials can also be influenced by the presence of microscopic defects and porosity. These mechanisms are primarily the result of a combined interaction between these phenomena and not a characteristic effect of each. Further, from Table 5, it is observed that the elastic modulus [E] of the nanocomposite sample increased with the addition of REO nanoparticles. This is consistent with the trend exhibited by magnesium (Mg)-based metal matrix composites [16] indicating a good interface between the reinforcing REO nanoparticles and the Mg matrix.

Conclusions

In this research study, pure magnesium (Mg) and Mg-1.5vol.% (Sm_2O_3 and CeO_2) nanocomposites were fabricated using microwave sintering followed by hot extrusion process. The microstructure, mechanical, thermal and damping performance were investigated. The key findings are the following:

(a) Scanning electron microscopy (SEM) results indicate that REO nanoparticles are homogeneously distributed throughout the Mg metal matrix.
(b) Compression tests revealed an increase in the strength from 249 ± 6 MPa for pure Mg to 395 ± 7 MPa for Mg-1.5 vol.% Sm_2O_3 and 329 ± 5 MPa for the Mg-1.5 vol.% CeO_2 nanocomposite.
(c) The observed increase or enhancement in the mechanical properties can be attributed to the conjoint influence of secondary processing, homogenously distributed particles and dispersion strengthening effect.
(d) Coefficient of thermal expansion showed a reverse trend indicating an increase in thermal stability.
(e) The ignition temperature of Mg-1.5 vol.% Sm_2O_3 showed the highest resistance to ignition (enhancement by ~69 °C) indicating superior thermal and dimensional stability.
(f) The Mg-1.5 vol.% CeO_2 nanocomposite showed the best damping (damping loss rate, damping capacity and elastic modulus) response.

The technique of microwave sintering has potentially wider range of advantages and applications in the preparation of magnesium (Mg)-based nanocomposites tools primarily because of the advantages associated with the heating mechanism.

References

1. Gupta M, Wong W (2015) Magnesium-based nanocomposites: Lightweight materials of the future. Mater Charact 105:30–46
2. Gupta M, Parande G, Manakari V (2017) An insight into high performance magnesium alloy/nano-metastable-syntactic composites. In: 17th Australian International Aerospace Congress: AIAC 2017, Engineers Australia, Royal Aeronautical Society, p 270
3. Manakari V, Parande G, Doddamani M, Gupta M (2017) Enhancing the ignition, hardness and compressive response of magnesium by reinforcing with hollow glass microballoons. Materials 10(9):997
4. Wakeel S, Manakari V, Parande G, Kujur MS, Gupta M (2018) Synthesis and mechanical response of NiTi SMA nanoparticle reinforced Mg composites synthesized through microwave sintering process. Mater Today: Proc 5(14):28203–28210
5. Parande G, Manakari V, Wakeel S, Kujur M, Gupta M (2018) Enhancing mechanical response of monolithic magnesium using Nano-NiTi (Nitinol) particles. Metals 8(12):1014
6. Parande G, Manakari V, Meenashisundaram GK, Gupta M (2017) Enhancing the tensile and ignition response of monolithic magnesium by reinforcing with silica nanoparticulates. J Mater Res 32(11):2169–2178
7. Czerwinski F (2015) The reactive element effect on high-temperature oxidation of magnesium. Int Mater Rev 60(5):264–296
8. Kujur MS, Mallick A, Manakari V, Parande G, Tun KS, Gupta M (2017) Significantly enhancing the ignition/compression/damping response of monolithic magnesium by addition of Sm2O3 nanoparticles. Metals 7(9):357
9. Eyring L (1979) The binary rare earth oxides, Handbook on the physics and chemistry of rare earths 3:337–399
10. Kujur MS, Mallick A, Manakari V, Parande G, Tun KS, Gupta MJM (2017) Significantly enhancing the ignition/compression/damping response of monolithic magnesium by addition of Sm_2O_3 nanoparticles 7(9):357
11. Kujur MS, Manakari V, Parande G, Tun KS, Mallick A, Gupta MJCI (2018) Enhancement of thermal, mechanical, ignition and damping response of magnesium using nano-ceria particles
12. Parande G, Manakari V, Meenashisundaram GK, Gupta M (2016) Enhancing the hardness/compression/damping response of magnesium by reinforcing with biocompatible silica nanoparticulates. Int J Mater Res 107(12):1091–1099
13. Czerwinski F (2014) Overcoming barriers, advanced materials & processes, 28
14. Ashraf PM, Shibli S (2007) Reinforcing aluminium with cerium oxide: A new and effective technique to prevent corrosion in marine environments. Electrochem Commun 9(3):443–448
15. Kim YM, Yim CD, Kim HS, You BS (2011) Key factor influencing the ignition resistance of magnesium alloys at elevated temperatures. Scripta Mater 65(11):958–961
16. Reddy MP, Manakari V, Parande G, Ubaid F, Shakoor R, Mohamed A, Gupta M (2018) Enhancing compressive, tensile, thermal and damping response of pure Al using BN nanoparticles. J Alloy Compd 762:398–408

Ordered Colloidal Crystals Fabrication and Studies on the Properties of Poly (Styrene–Butyl Acrylate–Acrylic Acid) and Polystyrene Latexes

I. H. Ifijen, E. U. Ikhuoria, S. O. Omorogbe and A. I. Aigbodion

Abstract Monodisperse poly(styrene–butyl acrylate–acrylic acid) (P(St–BA–AA)) and polystyrene (PS) colloidal suspensions were prepared using emulsion polymerization technique with a view to study their variation in properties. The P(St–BA–AA) latex had a slightly lower glass transition temperature (Tg) (106 °C) compared to the PS microspheres (104 °C). *TEM analysis revealed a core-shell morphology in the* P(St–BA–AA) latex particles, as compared to the PS particles. The synthesized latex suspensions were used to fabricate mono- and binary-sized colloidal crystals via the evaporation-induced self-assembly approach. SEM/AFM analyses showed spherically shaped particles that readily assemble into a closely-packed three dimensional highly periodic pattern with hexagonal symmetry. Also, the smaller sized P(St–BA–AA) particles (197.3 nm) in the binary crystals arranged themselves in a well-ordered manner around the larger sized PS particles (404.9 nm). The results showed that the modification made on the functional group of PS by polymerizing with butyl acrylate and acrylic acid produced P(St–BA–AA) latex with more improved properties compared to the as-synthesized PS.

Keywords Poly(styrene–butyl acrylate–acrylic acid) · Butyl acrylate · Acrylic acid · Colloidal crystals, binary colloidal crystal

Introduction

Polymeric colloidal spheres have been widely investigated as a prospect for the fabrication of colloidal crystal films for utilization in varieties of optical applications, such as bioassays, sensors, displays, colour, lasers and solar energy [1–3]. For the

I. H. Ifijen (✉) · S. O. Omorogbe · A. I. Aigbodion
Product Development Laboratory, Rubber Research Institute of Nigeria, P. M. B. 1049 Benin City, Nigeria
e-mail: larylans4u@yahoo.com

E. U. Ikhuoria (✉)
Department of Chemistry, University of Benin, Benin City, Nigeria
e-mail: esther.ikhuoria@uniben.edu

T. S. Srivatsan and M. Gupta (eds.), *Nanocomposites VI: Nanoscience and Nanotechnology in Advanced Composites*, The Minerals, Metals & Materials Series, https://doi.org/10.1007/978-3-030-35790-0_11

majority of these applications, narrow size distribution, size control, morphology, shape, copolymer composition, cost and surface electrical properties are of great importance [3–7]. In recent years, a myriad of procedures like emulsion, suspension, precipitation and dispersion polymerization has been demonstrated for synthesizing controlled polymeric microspheres with specific desire properties [3, 8–10] with emulsion polymerization among the most widely used.

Several techniques like gravity sedimentation, vertical deposition, spin coating, electrophoresis, among other techniques have been proven to be very useful for attaining self-assembly of polymer colloidal spheres with vertical deposition technique being the most extensively used. Among the myriad of polymers that have been previously used for the fabrications of colloidal crystals with compact periodic monodisperse colloidal particles, polystyrene (PS) is considered to be very valuable due to excellent properties like low water absorbing property, rigidity, low production cost, etc. [11–13]. Many practical applications have been achieved in the industry using this polymer [14, 15]. However, the presence of non-polar functional groups in its chemical structures can limit its usefulness in some practical applications that require the generation of photonic crystals with strong bonding strength with its substrate, among others. Depending on the nature of application that requires the utilization of polystyrene microspheres, functional groups can be incorporated into its structures to suit the required application. For example, Yao Meng et al. [16] generated colloidal crystal coloured films with improved mechanical properties and hydrophobicity from PS spheres by incorporating it with polyacrylate. Also, Jingxia Wang et al. [7] also improved the mechanical strength between the inter-phase of colloidal photonic crystals and its substrate by using hard core PS latex with elastomeric methyl acrylate and acrylic acid shell. Colloidal crystal films with a special FCC structure and tough mechanical strength were simply fabricated using latex spheres with a hard PS core and an elastometric PMMA/PAA shell. In an attempt to improve the properties of polystyrene, butyl acrylate and acrylic acid were incorporated into styrene monomers. This is the first time a correlative study on the influence of butyl acrylate (BA) and acrylic acid (AA) on PS in P(St–BA–AA) latex and that of the conventional PS latex has been studied.

In this study, the impact of incorporating butyl acrylate and acrylic acid into the structure of PS microspheres via emulsion polymerization was evaluated by characterizing the as-synthesized PS and P(St–BA–AA) latexes for the fabrication of mono-sized and binary-sized colloidal crystals.

Experimental

Materials

All reagents like butyl acrylate (BA), styrene, acrylic acid (AA), sodium hydroxide, acrylamide (AAm), sodium dodecyl benzenesulfonate (SDBS),

N, N-methylenebisacrylamide (99%) (BIS), ethylene glycol, N, N-methylenebisacrylamide (99%) (BIS), ammonium bicarbonate, (EG), potassium per-sulfate (KPS), pure nitrogen gas, ammonium per-sulphate (APS) and dimethoxy-acetophenone (DMAP) are of analytical grade.

Experimental Procedure

Surfactant-free emulsion polymerization method was used to prepare colloidal polystyrene (PS) spheres as described by Yohanala et al. [14]. Inhibitor present in the styrene was removed by washing with 1 M sodium hydroxide solution. Styrene (1.36 g) was transferred to a two-neck reaction flask containing 15 g of distilled water, and thereafter agitation was carried out with the aid of a magnetic stirrer (700 rpm). A temperature of 80 °C was maintained in the reaction vessel. The reaction was carried out in an inert atmosphere by continuously injecting nitrogen into the system. After 15 min, 0.05 g of potassium per-sulphate was introduced into the reaction mixture to initiate polymerization reaction. The reaction was stopped after 7 h. Centrifugation process was thereafter used to collect the prepared latex which was thoroughly washed with distilled water.

Preparation of Monodispersed P(St–BA–AA) Latex Spheres

Emulsion polymerization method was employed in the synthesis of poly (styrene–butyl acrylate–acrylic acid) [P(St–BA–AA)] colloidal latex as described by Minghui et al. [17] with modification. The styrene was washed with 1 M sodium hydroxide solution to remove the inhibitor present. Acrylic acid and Butyl acrylate were used without purification [14]. A double step polymerization method was preferred over the one-pot synthetic procedure described by Minghui et al. [17] because butyl acrylate undergoes polymerization in the presence of heat. In a routine experimental procedure, ammonium bicarbonate (0.085 g), acrylic acid (AA) (0.720 g), styrene (St) (3.190 g) and sodium dodecyl benzene sulphonate (SDBS) (0.005 g) were transferred into a 50 ml flask with distilled water (16.5 g). The mixture was agitated at 410 rpm with a magnetic stirrer in an inert environment for a period of about 30 min and temperature of 90°C. This was immediately followed by transferring a mixture of butyl acrylate (BA) (0.735 g) monomer and APS initiator (0.091 g) into the reaction flask. The reaction was allowed to stand for 13 h.

Characterization Techniques

Transmission electron micrographs of the PS and P(St–BA–AA) colloidal particles were viewed using FEI Tecnai 30 G2S-TWIN-TEM at an accelerating voltage of 300 kV. Atomic force micrographs were captured using Multi-Mode 8 AFM equipped with NanoScope V controller (Bruker, Santa Barbara, CA, USA). Scanning electron micrographs were taken using a scanning electron microscope (JEOL-JSM 5600 LV). Olympus BX51 polarized optical microscope was used to examine the morphology of the fabricated colloidal crystal films. Perkin-Elmer Series Spectrum Two FT-IR spectrometer was used to determine the functional in the polymer samples. PANalytical EMPYREAN instrument equipped with reference radiation of Cu Kα ($\lambda = 1.54$ Å) at an operating voltage of 45 kV was used to capture the powder X-ray diffraction pattern of both polymer samples. The thermal stability of the polymer samples was evaluated using a thermo-gravimetric analyzer TA Q50 at a heating rate of 10 °C/min under a nitrogen gas atmosphere. Dynamic light scattering (DLS) (Nano-Zetasizer, Malvern Instruments) was used to examine the polydispersity index (PDI) and particle diameters of the as-synthesized polymer samples at 25 °C under the scattering angle of 173° at 6333 nm wavelength. The glass transition temperature (Tg) was obtained using a DSC 2920 module in conjunction with the TA Instruments 5100 system at a scan rate of 10 °C/min under a nitrogen atmosphere.

Results and Discussion

Figure 1a shows bands of infrared spectra corresponding to the functional groups of PS and P(St–BA–AA) latexes. Some of the absorbance peaks in the prepared polymer samples were observed to be similar due to the existence of the polystyrene core in both as-synthesized polymeric samples. The peaks at 696 and 762 cm^{-1} were observed to be common to both prepared PS and P(St–BA–AA) samples and can be attributed to C–H out-of-plane bending vibration in methylene and benzene groups, respectively. A broad peak ranging from 1443 to 1505 cm^{-1} corresponding to aromatic C = C–C stretching vibration was observed in the spectrum of PS [18]. A similar absorbance peak was also seen in an equivalent position for P(St–BA–AA) latex, although the peak is split into two. The observed peak corresponding to 1599 cm^{-1} in the spectra of both PS and P(St–BA–AA) samples is due to aromatic C = C–C stretching vibration [18]. The similar peaks at 2915 cm^{-1} and 3037 cm^{-1} observed in the spectra of the synthesized samples are due to methylene and aromatic C–H groups, respectively [18]. The emergence of absorbance peaks at 1162 cm^{-1} and 1725 cm^{-1} in the spectrum of P(St–BA–AA) could be ascribed to C–O stretching vibration (ester bond) [19–21] and carboxylic acid (C = O stretch) [21–23], respectively. These peaks were not observed in the spectrum of PS. The presence of completely new peaks in the spectrum of P(St–BA–AA) as opposed to that of the as-synthesized PS sample could be due to the modification made on PS microspheres

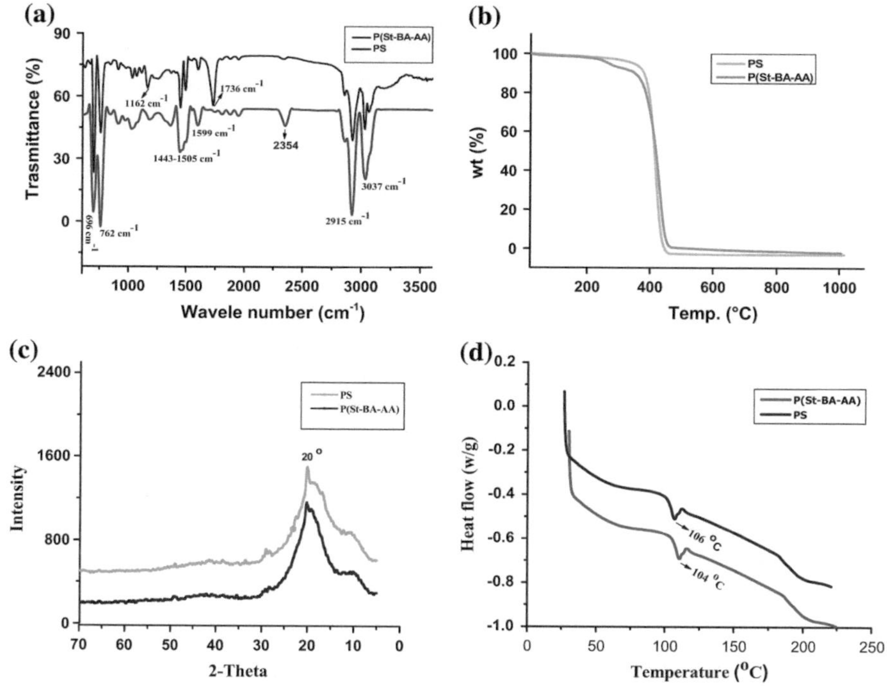

Fig. 1 **a** FTIR, **b** TGA, **c** XRD and **d** DSC patterns of PS and P(St–BA–AA) microspheres

during the synthetic stages of P(St–BA–AA), by incorporating butyl acrylate (BA) and acrylic acid (AA) into its structure.

Figure 1b shows the TGA results generated on samples of PS and P(St–BA–AA) at a heating rate of 10 °C/min. The spectrum of PS showed a fairly constant rate of degradation from 0 to 382 °C, whereas there was a slight decrease in the weight of P(St–BA–AA) between the temperature of 226 and 382 °C. This could be due to the loss of carbon (IV) oxide and water from the acrylate and ester portion of the terpolymer. As the temperature of 382 °C exceeds, there was a rapid decrease in the spectra of both samples until the temperature reached 382 °C. 14.2% loss in weight was observed during this period. Again, as the heating temperature increased beyond 382 °C, the rate of degradation for both samples further increased until the temperature reached 455 °C. The total weight loss during this period is 84.8%. The TGA analysis, therefore, showed that both samples were completely degraded at 455 °C.

Figure 1c revealed the XRD spectra of the as-synthesized PS and P(St–BA–AA) samples. Despite the fact that the chemical composition of both samples differs, equivalent absorption peaks with broad nature at position 2θ equal to 20° were observed for both samples. Both synthesized samples were concluded to be amorphous in nature because of the broad nature of their peaks. The obtained result is consistent with the observation of previous publications [23–25].

Figure 1d revealed the spectral of PS and P(St–BA–AA) obtained from DSC analysis. The glass transition temperatures (Tg) of the prepared samples were also evaluated because of the usefulness of thermal stability in determining feasible applications for polymer colloidal latexes. The acquired result sets the Tg of the as-synthesized PS and P(St–BA–AA) microspheres at about 106 °C and 104 °C, respectively. The results showed that the prepared P(St–BA–AA) sample has a slightly lower Tg (104 °C) when compared to the polystyrene latex (106 °C). The reduction in glass transition temperature from 106 to 104 °C may be due to the incorporation of the butyl acrylate (BA) unit to polystyrene chain during the formation of P(St–BA–AA) latex. The incorporation of BA to PS has been shown to decrease the Tg value of PS because polybutyl acrylate has a much lower glass transition temperature than polystyrene [26]. These temperature values indicate that the glassy and rigid nature of the synthesized PS and P(St–BA–AA) latexes will habitually be conserved whenever they are applied within 102 °C and 110 °C, respectively. However, their glassy and rigid nature becomes rubbery like or viscous when their glass transition temperature (Tg) is surpassed.

Table 1 shows the particle diameters, polydispersity indexes and zeta potentials of the synthesized PS and P(St–BA–AA) latexes. Average particle diameter, polydispersity index and zeta potential of about 197.5 nm, 0.051 and −31.90 mV were, respectively, obtained for PS and 404.9 nm, 0.027 and −37.40 mV for P(St–BA–AA) latex using dynamic light scattering (DLS) analysis (Table 1). The results showed that the synthesized PS and P(St–BA–AA) latex particles are monodispersed and of a good stable colloidal dispersion. However, the crystal films obtained using P(St–BA–AA) microspheres showed a slightly higher degree of monodispersity compared to the film obtained via self-assembly of PS.

Figure 2 shows the scheme of structure and transmission electron micrographs of as-prepared P(St–BA–AA) and PS latexes. The TEM micrograph of P(St–BA–AA) (Fig. 2b) revealed a distinct light and thick layer which signified a core-shell morphology. This observation was quite different from that of PS as no layer was observed in its TEM micrograph (PS) (Fig. 2c). The outermost layer (shell) of the latex sphere of the prepared P(St–BA–AA) latex (Fig. 2b) is made up of poly(butyl acrylate–acrylic acid) polar groups containing monomers (2a) while the inner layer (core) (Fig. 2b) is made up of non-polar monomer polystyrene group (PS) (Fig. 2a) [27].

Ordered microstructures were generated by subjecting the as-synthesized PS and P(St–BA–AA) latexes to vertical deposition technique at 55 °C to facilitate evaporation-induced self-assembly of the polymer colloidal particles. Figure 3 shows

Table 1 Particle diameter, polydispersity index and zeta potential of PS and P(St–BA–AA) latexes

	PS	P(St–BA–AA)
Average particle diameter (nm)	197.3	404.9
Polydispersity index (PDI)	0.051	0.027
Zeta potential (mV)	−31.90	−37.40

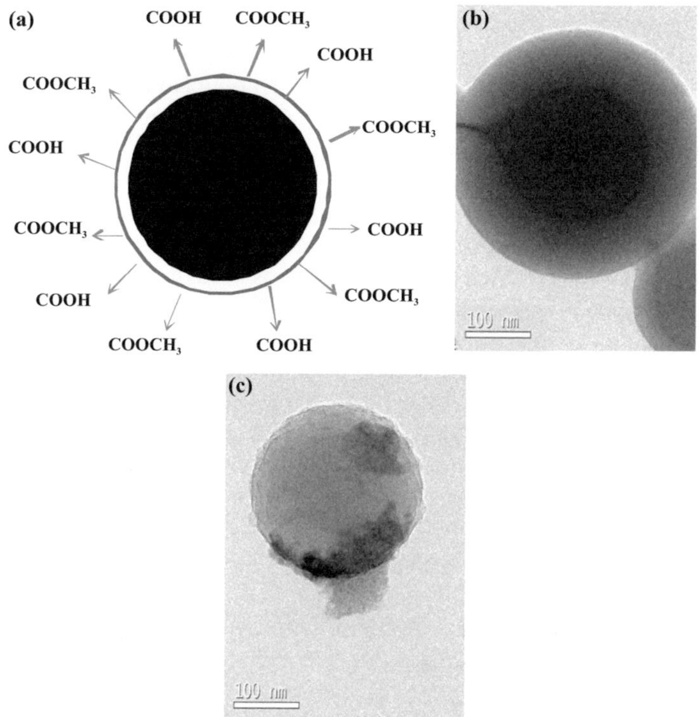

Fig. 2 **a** Scheme of the structure of the as-prepared P(St–BA–AA) latex sphere. **b** Typical TEM image of (**b**) core-shell P(St–BA–AA) latex spheres (**b**) PS

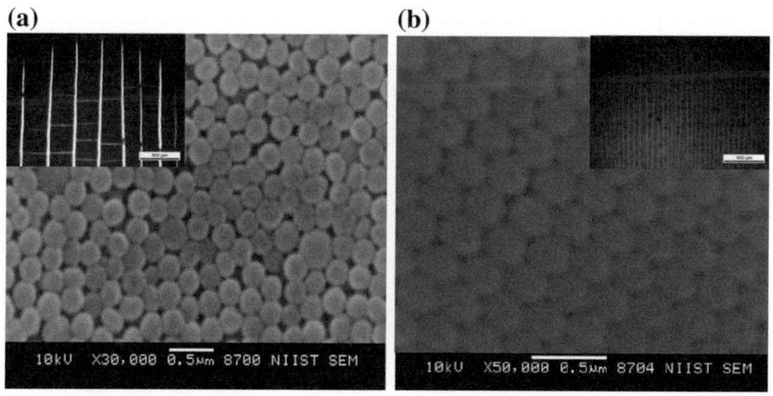

Fig. 3 Scanning electron micrographs of the prepared **a** PS and **b** P(St–BA–AA) with insets of their respective optical micrographs

the scanning electron micrographs of colloidal crystal films generated from the as-synthesized PS and P(St–BA–AA) colloidal particles with insets showing wider magnification images of the crystal film captured by an optical microscope. The results revealed periodic hexagonal alignment of latex spherical particles. However, P(St–BA–AA) particles (Fig. 3b) resulted in a slightly more compact morphology than the PS films (Fig. 3a). The arrangement of particles from both samples is in line with the polydispersity results shown in Table 1. The observed compact nature of the PS and P(St–BA–AA) crystal film particles has nothing to do with their chemical composition but may be attributed to the reaction conditions used during the synthetic stages of these polymer samples [21]. Polar carboxylic and acrylate groups have been shown by a previous study to have the capacity of improving the bonding strength existing between the interface of polymer colloidal particles and glass substrates [19]. As such, the bonding of the as-synthesized P(St–BA–AA) colloidal particles with glass substrates was expected to be higher than that of PS latex because of the presence of polar carboxylic and acrylate groups in its structure. The hydrophilic property of the as-synthesized P(St–BA–AA) latex may have also increased the hydrogen bond connections on the carboxylic (COOH) groups in the colloidal latex surface compared with the generated PS colloidal crystals [19]. The insets in Fig. 2a, b is a top surface view of an area in both colloidal crystal films viewed by a wide magnification optical microscope. The parallel lines revealed that there are few defects on the surfaces of the films. This could be attributed to solvent evaporation during the growth stage of the colloidal crystals.

Figure 4 shows three dimensional images (3D) of PS and P(St–BA–AA) colloidal crystal films. The AFM micrographs revealed uniform spherical particles that readily assemble into a highly periodic pattern with hexagonal symmetry. The observed morphology is in agreement with the morphology shown by the scanning electron micrographs (Fig. 3). The average particle diameter, particle height and surface roughness

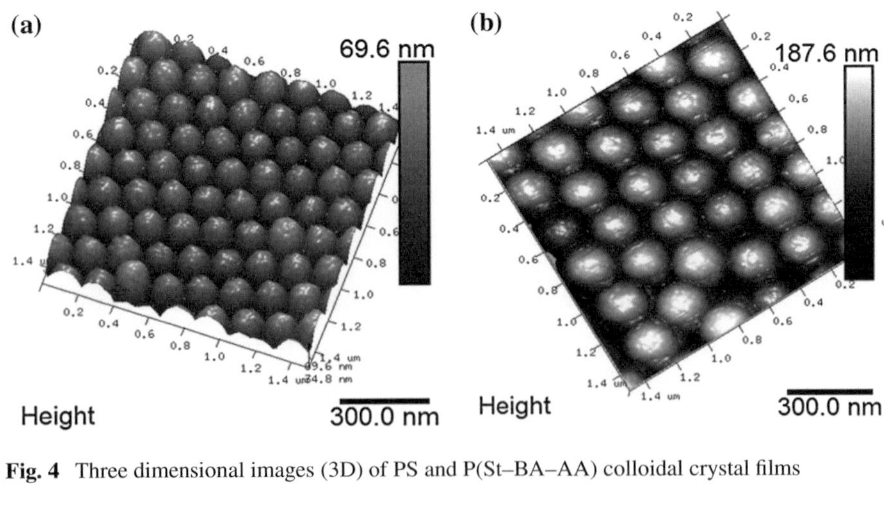

Fig. 4 Three dimensional images (3D) of PS and P(St–BA–AA) colloidal crystal films

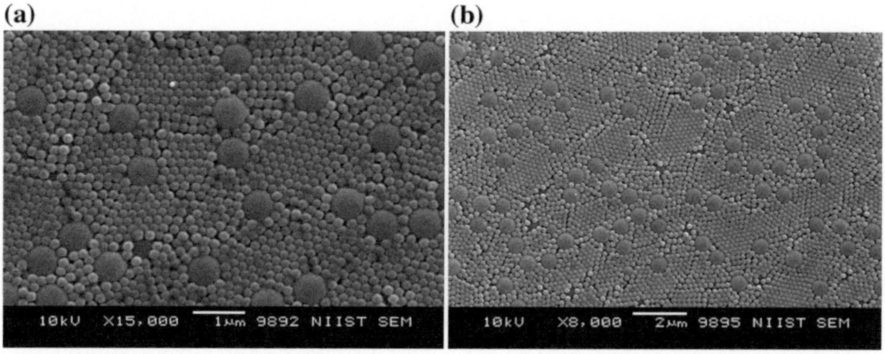

Fig. 5 Binary colloidal crystals fabricated from the as-synthesized PS and P(St–BA–AA) microspheres

of the fabricated films were estimated to be about 192.5 nm, 37 nm and 9.71 nm for PS and 401.6 nm, 41.5 nm and 9.61 nm for P(St–BA–AA), respectively.

Binary colloidal crystals have attracted great attention due to its fascinating potential for manipulating material properties towards utilization in, for instance, semiconductors, spintronics and photonics due to the fact that they permit the arrangements of particles with quite varied properties on a lattice [2].

Figure 5 shows binary colloidal crystal films fabricated using the as-synthesized PS and P(St–BA–AA) colloidal microspheres. Colloidal solutions of the PS (197.3 nm) and P(St–BA–AA) (404.9 nm) mixtures were prepared and thereafter subjected to vertical deposition procedure at 55 °C for the assembly of colloidal particles on the surface of a glass slide. SEM micrographs revealed that the colloidal particles readily assemble into an ordered periodic arrangement in such a way that the small size PS colloidal crystal spheres are assembled round the larger size P(St–BA–AA) particles in a well-ordered periodic manner. However, the numbers of small size particles were observed to be far more than the larger particles, and as such too many small particles were assembled around them. The formation of high-quality binary crystals may have been hindered by the fact that the colloidal particles used for this fabrication are micrometer-sized colloids as a previous study has established gravity and slow crystallization rates to be responsible for the hindered formation of micro-size high-quality binary crystals [2]. It could also be that the PS and P(St–BA–AA) mixing ratio used during the preparation of the colloidal particles did not meet the required mixing ratio that could generate binary colloidal crystals with the best possible qualities.

Conclusion

This study has demonstrated the synthesis of monodispersed (P(St–BA–AA) and PS latexes. Experimental results showed that the as-synthesized P(St–BA–AA) latex was more thermally stable than the prepared PS microspheres. Furthermore, spherical shaped particles that are readily assembled into a closely-packed three dimensional highly periodic pattern with hexagonal symmetry were generated using both colloidal latexes. Also, the fabrication of binary colloidal crystals with ordered particles was successfully carried out. The results disclosed that the improvement made on the functional groups of PS by incorporating with butyl acrylate (BA) and acrylic acid (AA) generated P(St–BA–AA) latex with a slightly more enhanced properties compared to the as-synthesized PS. This modified polystyrene may be useful in optical devices and decorative materials, such as high-performance reflectors, chemical sensors and environmentally friendly colour coatings.

Acknowledgements This investigation was supported by the World Academy of Sciences (TWAS). The authors also wish to appreciate Mr. and Mrs. Ifijen for their support.

Competing Interests The authors declare no conflicts of interest.

References

1. Arsenault AC, Puzzo DP, Manners I, Ozin GA (2007) Photonic crystal full color displays. Nat Photon 1(8):468–472
2. Vermolen ECM, Kuijk A, Filion LC, Hermes M, Thijssen JHJ, Dijkstra M et al (2009) Fabrication of large binary colloidal crystals with a NaCl structure. Proc Natl Acad Sci USA 106:16063–16067
3. Ríos-Osuna LA, Licea-Claverie A, Paraguay-Delgado F, Cortez-Lemus NA (2016) Synthesis of poly(styrene-acrylates-acrylic acid) microspheres and their chemical composition towards colloidal crystal films. Int J Polymer Sci 2016:10
4. Tian E, Wang J, Zheng Y, Song Y, Jiang L, Zhu D (2008) Colorful humidity sensitive photonic crystal hydrogel. J Mater Chem 18:1116–1122
5. Chung YW, Leu IC, Lee JH, Hon MH (2006) Influence of humidity on the fabrication of high-quality colloidal crystals via a capillary-enhanced process. Langmuir 22:6454–6460
6. Sheibat-Othman N, Burne T, Charcosset C, Fessi H (2008) Preparation of pH-sensitive particles by membrane contactor. Colloids Surf A 315:13–22
7. Wang J, Wen Y, Ge H, Sun Z, Zheng Y, Song Y et al (2006) Simple fabrication of full color colloidal crystal films with tough mechanical strength. Macromol Chem Phys 207:596–604
8. Xie C, Qiu T, Li J, Zhang H, Li X, Tuo X (2017) Nanoaramid dressed latex particles: the direct synthesis via pickering emulsion polymerization. Langmuir 33:8043–8051
9. Tan M, Shi Y, Fu Z, Yang W (2018) In situ synthesis of diblock copolymer nano-assemblies via dispersion RAFT polymerization induced self-assembly and Ag/copolymer composite nanoparticles thereof. Polymer Chem 9:1082–1094
10. Chen X, Ding Y, Ren D, Chen Z (2016) Green synthesis of polymeric microspheres that are monodisperse and superhydrophobic, via quiescent redox-initiated precipitation polymerization. RSC Adv 6:27846–27851
11. Yoon SB, Kim JY, Kim JH, Park SG, Kim JY, Lee CW et al (2006) Template synthesis of nanostructured silica with hollow core and mesoporous shell structures. Curr Appl Phys 6:1059–1063

12. León-Bermúdez A-Y, Salazar R (2008) Synthesis and characterization of the polystyrene—asphaltene graft copolymer by FTIR spectroscopy. CT&F—Ciencia, Tecnología y Futuro 3:157–167
13. Rouabah F, Dadache D, Haddaoui N (2012) Thermophysical and mechanical properties of polystyrene: influence of free quenching. ISRN Polymer Sci 2012:8
14. Yohanala F, Dewa RM, Quarta K, Widiyastuti, Winardi S (2015) Preparation of polystyrene spheres using surfactant-free emulsion polymerization. Modern Appl Sci 9:121–126
15. Berruezo M, Ludueña LN, Rodriguez E, Alvarez VA (2013) Preparation and characterization of polystyrene/starch blends for packaging applications. J Plast Film Sheeting 30:141–161
16. Meng Y, Tang B, Xiu J, Zheng X, Ma W, Ju B et al (2015) Simple fabrication of colloidal crystal structural color films with good mechanical stability and high hydrophobicity. Dyes Pigm 123:420–426
17. Wang M, Meng F, Wu H, Wang J (2016) Photonic crystals with an eye pattern similar to peacock tail feathers. Crystals 6:99
18. Fang J, Xuan Y, Li Q (2010) Preparation of polystyrene spheres in different particle sizes and assembly of the PS colloidal crystals
19. Duan G, Zhang C, Li A, Yang X, Lu L, Wang X (2008) Preparation and characterization of mesoporous zirconia made by using a poly (methyl methacrylate) template. Nanoscale Res Lett 3:118–122
20. Singho ND, Lah NAC, Johan MR, Ahmad R (2012) FTIR studies on silver-poly(Methylmethacrylate) nanocomposites via in-situ polymerization technique. Int J Electrochem Sci, 5596–603
21. Ifijen HI, Ikhuoria EU, Omorogbe SO (2019) Correlative studies on the fabrication of poly (styrene-methyl-methacrylate-acrylic acid) colloidal crystal films. J Dispers Sci Technol, 1–8
22. Hemalatha P, Veeraiah MK, Kumar SP, Madegowda NM, Manju M, Reactivity ratios of n-vinylpyrrolidone—acrylic acid copolymer. Am J Polymer Sci, 4:15–23
23. Ikhuoria EU, Omorogbe SO, Sone BT, Nuru ZY, Khamlich S, Maaza M (2018) Raspberry-like and other hexagonal close-packed morphologies of P(St-MMA-AA) particles obtained from different emulsifiers for photonic applications. J Mod Opt 65:1817–1826
24. Omorogbe SO, Ikhuoria EU, Ifijen HI, Simo A, Aigbodion A, Maaza M (2019) Fabrication of monodispersed needle-sized hollow core polystyrene microspheres. Springer International Publishing, Cham, pp 155–164
25. Ifijen HI, Ikhuoria EU, Omorogbe SO (2019) Correlative studies on the fabrication of poly(styrene-methyl-methacrylate-acrylic acid) colloidal crystal films. J Dispers Sci Technol, 1–8
26. Hiorns R (2000) In: Brandup J, Immergut EH, Grulke EA (eds), A Abe, DR Bloch (Assoc eds) Polymer handbook, 4th edn. Wiley, New York, (1999), pp 2250, price £210 ISBN 0-471-16628-6. Polym Int 49:807
27. Ifijen IH, Maliki M, Ovonramwen OB, Aigbodion AI, Ikhuoria EU (2019) Brilliant coloured monochromatic photonic crystals films generation from poly(styrene-butyl acrylate-acrylic acid) latex. J Appl Sci Environ Manage 23:1661–1664

Part IV
Polymer and Other Nanocomposites

Effect of Graphene Nanosheets Reinforcement on the Mechanical Properties of Rubber Seed Oil Based Polyurethane Nanocomposites

E. O. Obazee and F. E. Okieimen

Abstract Graphene-reinforced polyurethane nanocomposites were prepared by catalytic reaction of dispersion of exfoliated graphene nanosheet/rubber seed oil polyol (Gr–RSOP) hybrid and polyisocyanates (hexamethylene diisocyanate, HMDI, and 4,4'-methylene-bis(phenylisocyanate, MDI), at equimolar reactant ratios (NCO/OH ratio of 1.0), to give samples Gr–PUH and Gr–PUM, respectively. The structure and morphology of the obtained nanocomposites were analyzed using X-ray diffraction, atomic force microscopy (AFM) and FT–IR, respectively, while the mechanical and thermal properties were determined using nanoindenter, universal testing machine and thermogravimetric analyzer. The X-ray pattern revealed exfoliated graphene nanosheets in the nanocomposites, while the structures of the neat polyurethanes and nanocomposites showed great similarity. The hardness, tensile strength, young modulus, and thermal stability showed varied improvement and a corresponding reduction in elongation attributed to graphene incorporation.

Keywords Graphene · Rubber seed oil · Polyol · Polyurethane · Nanocomposites · Property

Introduction

The high crystalline nature of graphite hinders large polymer molecules from filling the inter-layer space of its unit graphene sheets, thus making it unsuitable in polymer reinforcement, and hence the need for its modification to graphene. This can run from expanding graphite (by several oxidizing agents) [1, 2] to its exfoliation. Notable exfoliation techniques are micromechanical exfoliation, liquid-phase exfoliation, epitaxial growth on SiC and chemical vapor deposition (CVD) growth

E. O. Obazee (✉)
Rubber Research Institute of Nigeria, P. M. B 1049 Benin City, Nigeria
e-mail: efosaosagie@yahoo.cco.uk

F. E. Okieimen
Department of Chemistry and Center for Biomaterials Research, University of Benin, Benin City, Nigeria

T. S. Srivatsan and M. Gupta (eds.), *Nanocomposites VI: Nanoscience and Nanotechnology in Advanced Composites*, The Minerals, Metals & Materials Series, https://doi.org/10.1007/978-3-030-35790-0_12

[3–9]. Graphene is a two-dimensional or single-layer graphite packed in a honeycomb lattice [9–13] with a thickness of one single atom [14], which is thought to be the stiffest [3, 15] and strongest material in the world [16]. Intrinsic graphene is a semi-metal or zero-based gap semi-conductor [9] with unique properties that vary significantly from the bulk graphite and have greatly influenced its incorporation in new technologies [17]. Pristine graphene has amazing mechanical, electrical, chemical and thermal properties (high thermal conductivity approx. 5000 $Wm^{-1} K^{-1}$), an ultra-high intrinsic carrier mobility (~200, 000 $cm^2V^{-1}S^{-1}$), large theoretical surface area (~2630 $m^2\ g^{-1}$), and high theoretical transmittance (97.7%). All these qualities of pristine graphene make it promise in many applications [9], such as in transparent conductive films, transistors, super capacitors, sensors, and catalyst support [14, 18–23].

A major encumbrance in the use of pristine graphene material is the difficulty in obtaining the unmodified form and insolubility in all organic solvents, as the modified graphene does not retain the structural quality and exceptional properties (e.g., as electron mobility) found in the single-layer graphene [18, 24]. Graphite oxidation to graphene oxide (GO), via modified Hummers method, and the reduction of the obtained GO to reduced graphene oxide (rGO) is the commonest method used to produce graphene, but with distorted structure that cannot be fully restored [24, 25]. Also, rGO is susceptible to aggregation and requires surfactant or strong capping agent to improve dispersion due to its hydrophobicity, and thereby introduces impurity and thus disrupts the electrical structure of pristine graphene as the surfactant is difficult to remove [26–29].

Without significant modification, exfoliation of graphene from graphite in the liquid phase can be achieved using both aqueous and non-aqueous systems, in reasonable quantities for research purpose that can be scaled up. Unlike the non-aqueous system, the aqueous system of liquid-phase exfoliation of graphene is cheap, safer and environmentally friendly from the use of water, only requiring the addition of surfactant in order to get the best interfacial tension range for efficient exfoliation. Matching the cohesive energy of liquid phase to the cohesive energy of graphite promotes the separation of graphitic and graphene sheets under shear [30]. Liu et al. [15] and co-worker demonstrated that graphite flakes can be exfoliated to give graphene sheets that are well dispersed in ethanol/water mixture and stable for years without addition of any surfactants.

Though there are several nanoparticles that are currently used in the synthesis of polymer nanocomposites (PCN), such as clays, carbon nanotubes, silica nanoparticles, nanofibers, etc., graphene is outstanding. Graphene sheets can be used in hydrophobic and hydrophilic polymers and can furnish most of the possible merit of nanorange sized fillers [31–33]. The low cost, natural availability, combined with its good mechanical, thermal, and barrier properties makes graphene a choice nanoparticle for use in polymer nanocomposites that can offer possibilities for new material development. Various methods that can be employed for the preparation of polymer graphene nanocomposites include solution blending [31, 34], exfoliation–adsorption [35], in situ intercalative polymerization [36] and melt intercalation [37].

There are some reports on expanded graphite or exfoliated GO, rGO, functionalized rGO, and nanocomposites based on a range of polymers [38–41] but there appear to be few reports on pristine graphene–polyurethane nanocomposites, and to the best of our knowledge, no report on rubber seed oil polyol based polyurethane nanocomposite. In this study, we report the preparation of rubber seed oil polyol (RSOP) based graphene-reinforced polyurethane nanocomposites.

Experimental

Materials and Methods

Mechanically expressed rubber seed oil (RSO) was obtained from Rubber Research Institute of Nigeria and used as received in the preparation of rubber seed oil polyol [42], physico-chemical properties given in Table 1. Laboratory grade graphite powder, −20–80 mesh, 99.9% (metal basis), was obtained from Alfa Aesar, USA. Hexamethylene diisocyanates (HMDI), 4,4’-methylene-bis(phenylisocyanate) (MDI), toluene, ethanol, dichloromethane, and dibutyltindilaurate (DBTDL) were obtained from Sigma Aldrich, India.

Table 1 Physico-chemical properties of RSOP

Properties	RSOP
Color	Orange-yellow
Density	1.019
AV ($mgKOHg^{-1}$)	14.733
SV ($mgKOHg^{-1}$)	239.17
IV ($gI_2 100\ g^{-1}$)	9.240
HV ($mgKOHg^{-1}$)	203.47
Oxirane content (%)	0.402
Mn ($g\ mol^{-1}$)	1014.68
Functionality	3.7524

Characterization Techniques

Structural and Morphological Analyses

Structural studies of the Polyurethane and nanocomposites samples were recorded using Agilent Technologies Cary 660 FTIR spectroscopy attached with attenuated total reflectance (ATR). Rigaku Smart lab Wide Angle X-ray Diffraction (WAXD) machine using CuKa radiation (at $\lambda = 1.54068$ A°) and a Bragg–Brentano geometry, equipped with X-ray generator set at 45kV tension and 100 mA current, a point proportional detector, and a curve quartz monochromator, a goniometer radius circle of 250 mm were used for examining the crystalline and morphology of the samples. A scanning rate of 0.02°/s over range of 2θ (2–50°) for one-dimensional X-ray diffraction (XRD) pattern on oriental samples suspended on a glass slide was used for the X-ray studies. Also, the Bruker Dimension ICON ScanAsyst Atomic Force Microscopy (AFM) was used for the surface morphology, using the tapping mode.

Physico-Mechanical Properties of Polyurethanes and Nanocomposites

Tinius Olsen H50KS Universal Testing Machine (using ASTM 6287-09 method, 50KN force at a speed of 50 mm/min and gauge length of 12.04) and Hystron TI 950 Triboindenter, with a standard diamond indenter probe, using ASTM E2546-07, were used for determining the tensile properties and hardness, respectively. The thermal stability of the samples was carried out with the Netzsch STA 449 FI Jupiter analyzer at a heating rate of 10 °C/min, to a temperature of 600 °C, using sample weight between 10 and 14 mg at 10 k/min in nitrogen. The degree of swelling of the polymers was carried out by immersing cut-and-weighed size pieces into a solution of toluene and taking the weight at interval at ambient temperature until a constant weight was obtained. At this point, the sample was blotted dry with a dry cloth and weighed. Duplicate experiment was carried out for each sample and the average was taken as the final weight [43].

Preparation of Graphene-Reinforced Polyurethane Nanocomposites

Exfoliation of Graphene Nanosheets

The method of Liu et al. [15]was used in the preparation of the mono- or few layered graphene (Gr), by sonication of graphite powder in ethanol/water solution. In a typical experiment, 1 g of graphite powder, −20–80 mesh and 99.9% metal basis, was weighed into a vial, and 40 ml solution of 70/30 ethanol/water solution was poured in. The mixture was shaken for about a minute and placed in an ultrasonic

Fig. 1 **a** Centrifuged and sonicated graphite/graphene dispersion in ethanol–water. **b** RSOP. **c** Pipetted upper layer supernatant graphene dispersion (middle vial is ethanol–water reference). **d** Mixture of pipette graphene layer in ethanol–water and RSOP

water bath, at 25 °C, and dispersed for about 3 h, after which it was centrifuged for 60 min at 500 rpm. The supernatant upper layer graphene was pipetted off, using a micro-pipette, into a pre-weighed reactor flask, see Fig. 1a, c.

Preparation of Graphene–Rubber Seed Oil Polyol (RSOP) Hybrid

A calculated amount of rubber seed oil polyol (RSOP) (required in preparing polyurethanes at NCO/OH ratio 1.0) was added to the flask containing the as-obtained graphene dispersed in ethanol/water described above. The flask was swirled to dissolve the RSOP and to ensure proper mixing and placed in a rotary evaporator to remove the ethanol/water solution, first at 40 °C, and then at 80 °C, in a total of about 24 h, until the graphene–rubber seed oil polyol (Gr-RSOP) hybrid so obtained was solvent and moisture free.

Preparation of Polyurethanes and Graphene-Reinforced Polyurethane Nanocomposites

Into two quick-fit flasks containing Gr-RSOP hybrid obtained above and RSOP, respectively, under reflux and fitted with nitrogen gas inlet and was charged the required amount of DBTDL catalyst and toluene, and stirred to enable proper mixing in an inert atmosphere. Then the calculated amount of HMDI required to give an NCO/OH ratio of 1.0 was calculated (see recipe on Table 2), and was charged into the

Table 2 Recipe of graphene-reinforced polyurethanes nanocomposites

Samples	Ingredients	Graphene content (%)
PUH	RSOP—HMDI (NCO/OH ratio 1.0)	0
Gr–PUH	RSOP—HMDI (NCO/OH ratio 1.0)	0.13
PUM	RSOP—MDI (NCO/OH ratio 1.0)	0
Gr–PUM	RSOP—MDI (NCO/OH ratio 1.0)	0.13

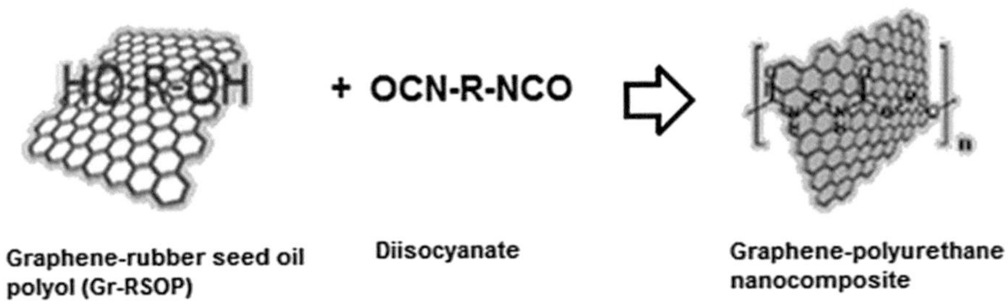

Fig. 2 Preparation of graphene-reinforced polyurethane nanocomposites

flasks, and the mixture was stirred and de-gassed, and then poured into a pre-greased mold to give graphene-reinforced polyurethane (Gr–PUH) and neat polyurethane (PUH) samples. The samples (Gr–PUH and PUH) were cured for about 2 h at 60 °C, and finally for about 12 h at 100 °C. This process was repeated using MDI as the polyisocyanate to give Gr–PUM and PUM samples. Figure 2 is a schematic representation of the reaction processes. The graphene loading per nanocomposites prepared was about 0.13%, based on the content of polyol.

Results and Discussion

Morphology of polyurethane nanocomposites

Insight into the degree of nanodispersion of the polyurethane nanocomposites was evaluated with WAXD. From the XRD diffraction patterns of the nanocomposite and neat polyurethanes (Fig. 3), the amorphous nature of the polyurethane nanocomposites (Gr–PUH and Gr–PUM) with little crystallinity, similar to the neat polyurethanes (PUH and PUM), is revealed. Thus, the molecules of the polyurethanes and nanocomposites are mainly irregularly packed with disordered orientation. This is due to the fact that polymers are composed of low atomic number elements with small atomic scattering factors and have only small content of crystal present [38]. For comparison, the XRD pattern of graphite (G), graphene oxide (GO), graphene nanosheet

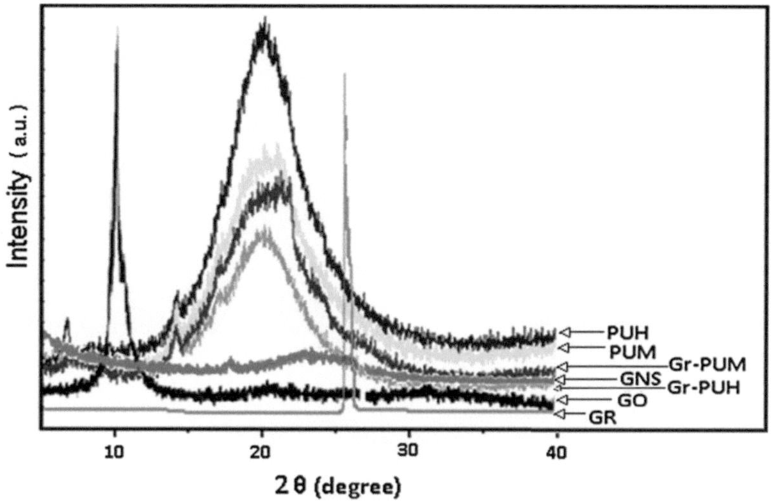

Fig. 3 X-ray diffractions of graphite, graphene nanosheets, graphene oxide, HMDI and MDI polyurethanes and their graphene-reinforced nanocomposites

(GNS), and Gr-PUs are shown with their characteristic peaks. The absence of the sharp planar (hexagonal arrangement of carbons) peak of graphite at about 27.4° in the 2θ degree of the Gr–PU XRD is an indication of the absence of graphite. Suffice to say also that it affirms to the fact that graphite was not oxidized to its oxide (GO) in the processing stage, nor did it remains in the graphite state. This could only point to the fact that there was successful dispersion (exfoliation) of graphite in ethanol/water mixture to single and few layered graphene nanosheets according to the technique reported by Liu et al. [15], neither did it restack or agglomerate back to graphite in the course of its dispersion into RSOP, nor in the preparation of the polyurethane nanocomposites. This is further confirmed by the XRD of the exfoliated graphene, which shows no prominent peak, except a very little hump (23–27°) subsumed in the halo peaks of the polyurethanes. And there are no other visible peaks in this region in the XRD of the nanocomposites that suggest otherwise.

The tapping mode AFM technique was also used in the morphological study of the samples. In Fig. 4a, b are shown the soft (dark) and hard (bright) domain segments of pristine polyurethanes elastomeric PUH and the rigid PUM surfaces, which are typical with polyurethanes. Detailed information on the nature and dimension of the dispersed phase on the continuous matrix phase, as seen in Fig. 4c, was not possible due to confusion with the characteristic segmentation domains of the polyurethanes and absence of SEM and TEM analysis of samples. However, the surface roughness as average Root Mean Square (Rq) values of Gr–PUH (10.13 nm) and Gr–PUM1 (21.8 nm) which are higher than those of the pristine polyurethanes (PUH and PUM), with an average Root Mean Square (Rq) values of 7.125 nm and

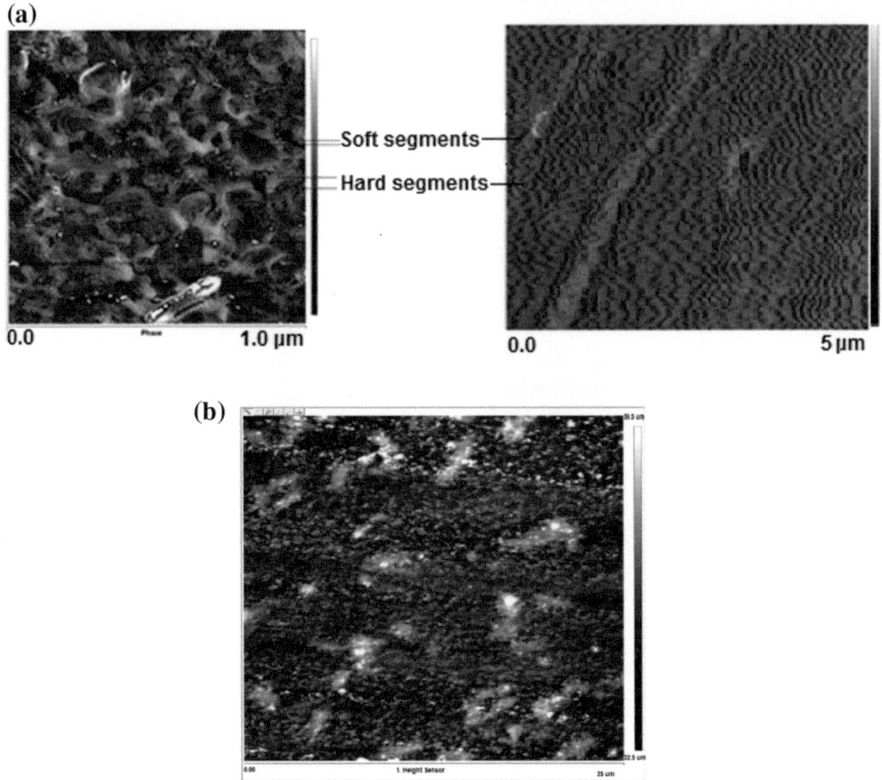

Fig. 4 **a** AFM of PUH and PUM showing the segment, **b** AFM of Gr–PUH

19.25 nm, respectively, was obtained with the AFM. This increase in surface roughness of the graphene-reinforced polyurethane nanocomposites could be attributed to the inclusion of the graphene nanosheet fillers.

The FTIR is a vital instrument used in the structural elucidation of polyurethane, as it gives very vital information of two major band regions, namely, the urethane linkage-NH region and the carbonyl ($-C = O$) region of the polymer. The FTIR spectra of the polyurethanes and their nanocomposites (Gr–PUH and Gr–PUM) are shown in Fig. 5a, b and are observed to be very similar with their respective pristine polymer. The spectra show the typical donor N-H and acceptor $C = O$ stretching absorption bands of polyurethane at 3300–3360 cm^{-1} and 16950–1735 cm^{-1}, respectively, the asymmetric and symmetric stretching absorption bands of CH (2865–297) and present in the MDI derivatives are the $C = C$ in-plane aromatic peaks. In the carbonyl region, the presence of the associated and non-associated stretching bands, which represents the level of H-bonding and free or non-hydrogen bonding within the

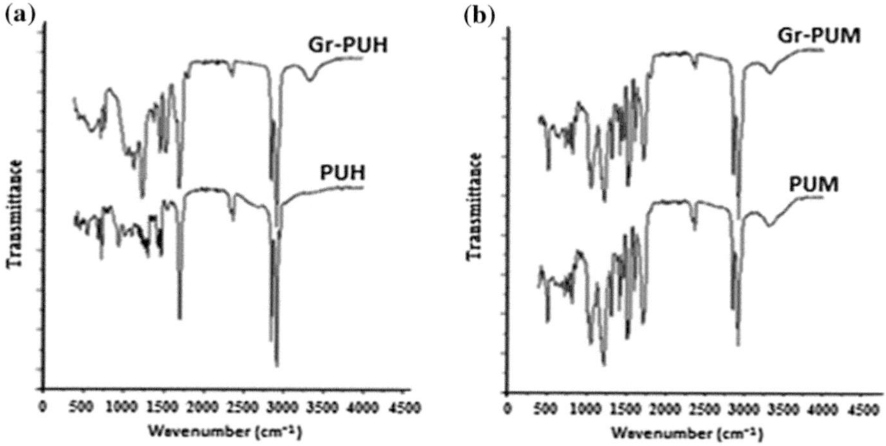

Fig. 5 FTIR of **a** HMDI polyurethane, PUH, and graphene polyurethane nanocomposite (Gr–PUH). **b** IR of MDI polyurethane (PUM) and graphene polyurethane nanocomposite (GrPUM)

urethanes structures, is visible. It is observed that the peaks of the H-bonded C = O are more intense and broad, which also reflects the level of inter-molecular activities in the network. There is an exceptional little shoulder around 1785–1800 cm^{-1}, present in the two nanocomposites spectra, absent in those of the pristine polyurethanes.

Mechanical Properties

Pristine graphene is reported to have the highest theoretical strength of all recorded known materials [44], and it is expected that this will greatly impart on physico-mechanical properties of the polyurethanes, such as hardness (ability to resist indentation), strength and modulus, thermal stability, etc., where there is proper mixing and exfoliation (Table 3).

Table 3 Physico-mechanical properties of polymer and nanocomposites

Sample	Young's modulus (GPa)	Average hardness (Mpa)	Tensile strength (Mpa)	Total elongation (MPa)	Ultimate true stress (MPa)	Yield stress (MPa)	Swelling @30° C (%)
PUH	0.016 ± 001	1.399 ± 001	2.09	187	5.93	2.09	1. 58
Gr–PUH	0.020 ± 002	1.580 ± 003	2.12	83.9	3.92	1.75	1.54
PUM	1.963 ± 001	60.50 ± 001	5.17	161	12.6	5.17	1.51
Gr–PUM	2.967 ± 001	136.92 ± 001	8.56	113	9.08	7.68	1.46

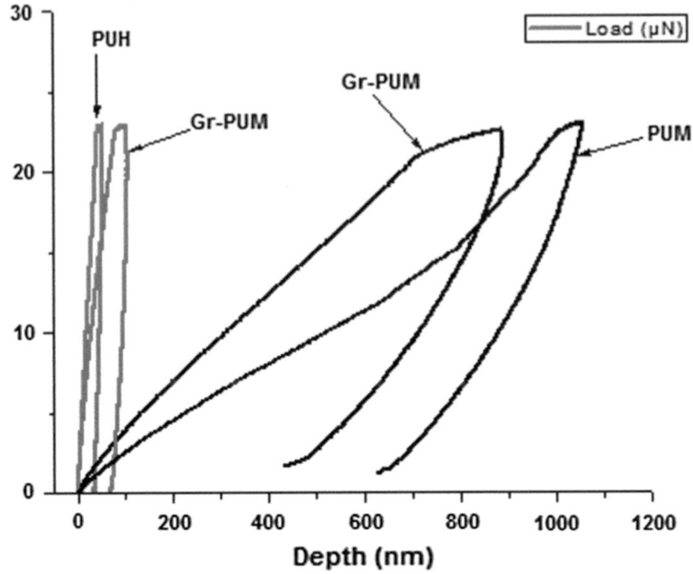

Fig. 6 Loading–unloading curves of polyurethanes and nanocomposites

Hardness

The loading–unloading curves of the polyurethanes (PUH and PUM), which are soft and rigid, respectively, and the nanocomposites (Gr–PUH and Gr–PUM), which are both rigid, are shown in Fig. 6 GR–PUH and Gr–PUM with about 12.9% and 126.2%, respectively, from the pristine PUH and PUM. The relatively high level of improvement recorded in Gr–PUM compared to Gr–PUH could be due to the combinative effect of graphene's hardness, brought to fore by good dispersion and interaction at the interphase, and the rigid nature of the pristine aromatic polyurethane facilitated by Van der Waals forces and hydrogen-bond interactions [44].

Tensile Strength and Modulus

The stress–strain curves of PH, PUM, Gr–PUH and Gr–PUM are shown in Fig. 7, and it is seen that there is an increase in the tensile strength of the nanocomposites relative to the unfilled polyurethanes, which led to 1.4 and 65.57% increase in Gr–PUH and Gr–PUM, respectively. The comparatively lesser improvement in Gr–PUH could be due to the inability of the graphene nanosheet and the polyurethane matrix

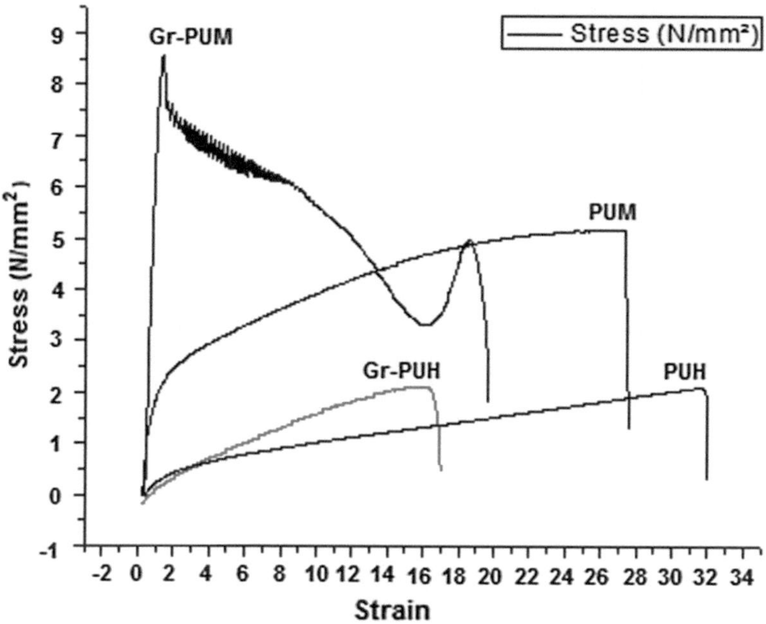

Fig. 7 Stress–strain curves of polyurethanes and nanocomposites

to have an optimized intereactivity at the interphases, resulting in the very low force of intimacy or homogeneity. This explanation is supported by the drastic reduction in its elongation at break, which is 83.9% (about 55.1% reduction from PUH), however. There was no sign of graphene stacking, as evidenced in the lack of the graphitic peak in the Gr–PUs X-ray diffractions. The relatively high increment in Gr–PUM could be due to relatively good dispersion and interaction at the interphases of the graphene sheets and polymer matrix, in addition to the rigid nature of the aromatic diisocyanate based polyurethane, resulting from higher H-bonding content, the effect of the aromatic ring, and higher crosslink in the network.

The Young's modulus is observed to increase by 25 and 51.1% in Gr–PUH and Gr–PUM. However, these increments in hardness and strength were followed by a corresponding decrease in elongation at break. As stated earlier, the unprecedented reduction in the elongation in Gr–PUH could be due to the failure at the interfacial region which did not have optimum interaction at the interphase of the graphene nanosheet and matrix of the polymer. Also, there is a reduction in the ultimate true stress in Gr–PUH and Gr–PUM, while yield stress increased in Gr–PUM and decreased in Gr–PUH. There is generally some improvement in the mechanical property of the polyurethane nancomposites (see Fig. 8), which could be ascribed to the impact of the presence of graphene reinforcement.

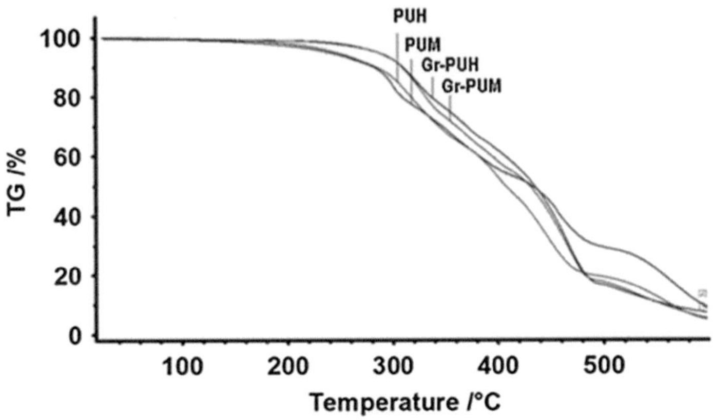

Fig. 8 Thermograms of polyurethanes and nanocomposites

Thermal Stability

The TGA results of the nanocomposites are listed in Table 4, while the respective thermograms are shown in Fig. 8.

The onset of degradation was at 250 °C for Gr–PUH and Gr–PUM each, representing an increment of about 23.7% and 219%, respectively, compared to the pristine polyurethanes. There is an increase in the temperature of the first step of degradation from 298°C to 297°C in PUH and PUM, to 310 °C, in Gr–PUH and Gr–PUM, i.e., about 4.0% and 4.1% increase, respectively. However, it was observed that unlike the pristine PUH, Gr–PUH thermogram had three steps, with the middle step not as pronounced, showing some level of stability at about 419 °C, which could be ascribed to the aromatic rings present in graphene.

This same step is seen in the thermograms of PUM and Gr–PUM, which is the middle or second degradation step, also due to degradation of the aromatic ring from the MDI aromatic ring, occurring at temperature of 430 °C.

Table 4 Thermal properties of Polyurethanes

Sample	Onset (°C)	First step (°C)	Second step (°C)	Third step (°C)
PUH1	202	298	515	–
Gr–PUH	255	310	420	530
PUM1	205	297	430	520
Gr–PUM	255	310	430	510

Conclusions

1. Graphene (mono- and few layered) sheets prepared by the method of Liu et al. [15], via aqueous liquid-phase technique, were successfully dispersed in RSOP co-monomer without re-aggregation, and used to prepare RSO polyol based graphene–polyurethane nanocomposites.
2. The inclusion of graphene brought about improvement in some of the properties of the nanocomposites.
3. It is quite impressive to note that at filler level of about 0.13% there is improvement in some properties of polyurethane nanocomposites.
4. This inclusion must satisfy certain conditions, such as good dispersion and/or optimized interactions at the interphases, for significant improvement to be obtained, as seen in Gr–PUH and Gr–PUM.

Acknowledgements The authors wish to acknowledge with gratitude Prof. Timothy Gonsalves, the Director of Indian Institute of Technology Mandi, Himachal Pradesh, India, for the Research Internship granted to EOO that made this possible, and Prof. I.O. Eguavoen, the Executive Director of Rubber Research Institute of Nigeria for the research leave granted to EOO.

References

1. Inagaki M, Tashiro R, Washino Y, Toyoda M (2004) J Phys Chem Solids 65:133
2. Chung DDL (1987) J Mater Sci 22:4190
3. Novoselov KS, Geim AK, Morozov SV, Jiang D, Zhang Y, Dubonos SV, Grigorieva IV, Firsov AA (2004) electric field effect in atomically thin carbon films. Science 306(5696):666–669
4. Nandamuri G, Roumimov S, Solanki R (2010) Chemical vapor deposition of graphene films. Nanotechnology 21–145604
5. Bae S, Kim H, Lee Y, Xu X, Park JS, Zheng Y, Balakrishnan J, Lei T, Ri KH, Song YI, Kim YJ, Kim KS, Ozyilmaz B, Ahn JH, Hong BH, Iijima S (2010) Roll-to-roll production of 30-inch graphene films for transparent electrodes. Nat Nano 5(8):574–578
6. Shivaraman S, Barton RA, Yu X, Alden J, Herman L, Chandrashekhar MVS, Park J, Mc Euen PL, Parpia JM, Craighead HG, Spencer MG (2009) Free-standing epitaxial graphene. Nano Lett 9(9):3100–3105
7. Aristov VY, Urbanik G, Kummer K, Vyalikh DV, Molodtsova OV, Preobrajenski AB, Zakharov AA, Hess C, Hänke T, Büchner B, Vobornik I, Fujii J, Panaccione G, Ossipyan YA, Knupfer M (2010) Graphene synthesis on cubic sic/si wafers. Perspectives for mass production of graphene-based electronic devices. Nano Lett 10(3):992–995
8. Emtsev KV, Bostwick A, Horn K, Jobst J, Kellogg GL, Ley L, Mc Chesney JL, Ohta T, Reshanov SA, Rohrl J, Rotenberg E, Schmid AK, Waldmann D, Weber HB, Seyller T (2009) Towards wafer-size graphene layers by atmospheric pressure graphitization of silicon carbide. Nat Mater 8(3):203–207
9. Deng D, Pan X, Zhang H, Fu Q, Tan D, Bao X (2010) Frees tanding graphene by thermal splitting of silicon carbide granules. Adv Mater 22(19):2168–2171
10. Cui X, Zhang C, Hao R, Hou Y (2011) Liquid-phase exfoliation, functionalization and application of graphene, nanoscale 3:2118–2126
11. Mittal V (2012) Polymer–Graphene nanocomposites. RSC nanoscience & nanotechnology No. 26. The Royal Society of Chemistry. Published by the Royal Society of Chemistry, www.rsc.org

12. Allen ML, Tung VC, Kaner RB (2010) honeycomb carbon: a review of graphene. Chem Rev 2010(110):132–145
13. Chen Z, Chisholm B, Patani R, Wu J, Fernando S, Jogodzinski K, Webster D (2010) Soybased UV-curable thiol–ene coatings. J Coat Technol Res 7:603–613
14. Xu M, Zhang W, Yang Z, Yu F, Ma Y, Hu N, He D, Liang Q, Su Y, Zhang Y (2015) One-pot liquid-phase exfoliation from graphite to graphene with carbon quantum dots. Nanoscale 7:10527–10534
15. Liu W, Bao-Yu Xia B-Y, Xiao-Xia Wang X-X, Wang J-N (2012) Exfoliation and dispersion of graphene in ethanol-water mixtures. Front Mater Sci 6(2):176–182
16. Meyer JC, Geim AK, Katsnelson MI, Novoselov KS, Booth TJ, Roth S (2007) Nature 446:60
17. Lee C, Wei X, Kysar JW, Hone J (2008) Science 321:385
18. Sham AYW, Notley SM (2013) A review of fundamental properties and applications of polymer–graphene hybrid materials. Soft Matter 9:6645–6653
19. Zhang YB, Small JP, Amori MES, Kim P (2005) Phys Rev Lett 94:176–803
20. Su CY, Xu YP, Zhang WJ et al (2009) Electrical and spectroscopic characterizations of ultra-large reduced graphene oxide monolayers. Chem Mater 21(23):5674–5680
21. Stoller MD, Park S, Zhu Y et al (2008) Graphene-based ultracapacitors. Nano Lett 8(10):3498–3502
22. Lin YM, Jenkins KA, Valdes-Garcia A et al (2009) Operation of graphene transistors at gigahertz frequencies. Nano Lett 9(1):422–426
23. Si YC, Samulski ET (2008) Exfoliated graphene separated by platinum nanoparticles. Chem Mater 20(21):679–6792
24. Skaltsas T, Karousis N, Yan H-J, Wang C-R, Pispas S, Tagmatarchis N (2012) Graphene exfoliation in organic solvents and switching solubility in aqueous media with the aid of amphiphilic bock copolymers. J Mater Chem 2012(22):21507
25. Gómez-Navarro C, Weitz RT, Bittner AM et al (2007) Electronic transport properties of individual chemically reduced graphene oxide sheets. Nano Lett 7(11):3499–3503
26. Dimiev A, Kosynkin DV, Alemany LB et al (2012) Pristine graphite oxide. J Am Chem Soc 134(5):2815–2822
27. Tung VC, Allen MJ, Yang Y et al (2009) High-throughput solution processing of large-scale graphene. Nat Nanotechnol 4(1):25–29
28. Stankovich S, Dikin DA, Dommett GHB, Kohlhaas KM, Zimney EJ, Stach EA, Piner RD, Nguyen ST, Ruoff RS (2006) Nature 442(7100):282–286
29. Luo Z, Lu Y, Somers LA et al (2009) High yield preparation of macroscopic graphene oxide membranes. J Am Chem Soc 131(3):898–899
30. Tang LH, Wang Y, Li YM et al (2009) Preparation, structure, and electrochemical properties of reduced graphene sheet films. Adv Func Mater 19(17):2782–2789
31. Notley SM (2012) Langmuir 28:14110–14113
32. Zheng W, Wong S-C (2003) Compos. Sci. Technol. 63(2):225–235
33. Zheng W, Wong S-C, Sue H-J (2002) Polymer 73(25):6767–6773
34. Xu J, Hu Y, Song L, Wang Q, Fan W, Liao G, Chen Z (2001) Polym Degrad Stab 73(1):29–31
35. Sadasivuni KK, Ponnamma D, Kim J, Thomas S (eds) (2015) Graphene-based polymer nanocomposite in electronics, VI, 382 p 175. ISBN 978-3-319-13874-9
36. Jing-Wei S, Xiao-Mei C, Wen-Yi H (2003) J Appl Polym Sci 88(7):1864–1869
37. Liu PG, Xiao P, Xiao M, Gong K-C (2000) Chin J Polym Sci 18(5):413–418. Wenge Z, Xuehong L, Shing-Chung WJ (2004) Appl Polym Sci, 91(5):2781–2788
38. Kim H, Abdala AA, Macosko CW (2010) Graphen/polymer nanocomposites. Macromolecules 43:6515–6530
39. Kuilla T, Bhadra S, Yao D, Kim NH, Bose S, Lee JH (2010) Recent advances in graphene based polymer composites. Prog Mater Sci 35:1350–1375
40. Yang D, Velamakanni A, Bozoklu G et al (2009) Chemical analysis of graphene oxide films after heat and chemical treatments by X-ray photoelectron and Micro-Raman spectroscopy. Carbon 47(1):145–152

41. Liu S, Tian M, Yan B, Yao Y, Zhang L, Nishi T (2015) High performance dielectric elastomers by partially reduced graphene oxide and disruption of hydrogen bonding of polyurethanes. Polymer 56:375–384
42. Obazee EO (2018) PhD Thesis. Biobased polymers from modified rubber seed oil, University of Benin City, Benin City
43. Ferrer CC, Babb D, Ryan AJ (2008) Characterization of polyurethane networks based on vegetable derived polyol. Polymer 49:3279–3287
44. Singh V, Joung D, Zhai L, Das S, Khondaker SI, Seal S (2011) Graphene based materials: past, present and future. Prog Mater Sci 56:1178–1271

Continuous Flow Process for Removal and Recovery of Water Contaminants with Magnetic Nanocomposites

Teagan Leitzke, Jerome P. Downey, David Hutchins and Brian St. Clair

Abstract Many natural water sources and industrial wastewaters contain low concentrations of metals and other contaminants. Therefore, an efficient and economical method for low-level contaminant removal and recovery is needed. The purpose of the research is to improve and modify a newly developed continuous flow ion exchange process for expansion to a variety of non-industrial applications, including removal of metal ions from the Upper Clark Fork River Watershed. The process involves a dual column reactor designed to capture metal ions using 90–105 μm spherical, functionalized silica gel coated magnetite particles, targeting copper ions with future expansion to additional metals, such as manganese and zinc. The optimization of nanoparticle synthesis and dispersion is ongoing with variables that include pH, metal ion concentration, nanoparticle concentration, and temperature. Additional focus involves maximizing contaminant capture, with current values of 0.19 mmol Cu/g Fe_3O_4 for magnetite and 0.25 mmol Cu/g Fe_3O_4 for silica-coated magnetite.

Keywords Magnetite nanoparticles · Nanocomposites · Wastewater treatment · Ion exchange · Adsorption

Introduction

Waste contamination can leave unacceptable concentrations of heavy metals in water sources, creating environmental and health concerns. The U.S. Environmental Protection Agency (EPA) and Montana Department of Environmental Quality (DEQ) have developed water standards to address these concerns. Three major elements often found in contaminated mine wastewaters are copper, manganese, and zinc,

T. Leitzke (✉) · J. P. Downey (✉) · D. Hutchins · B. St. Clair
Montana Technological University, 1300 W. Park St., Butte, MT 59701, USA
e-mail: tleitzke@mtech.edu

J. P. Downey
e-mail: jdowney@mtech.edu

T. S. Srivatsan and M. Gupta (eds.), *Nanocomposites VI: Nanoscience and Nanotechnology in Advanced Composites*, The Minerals, Metals & Materials Series, https://doi.org/10.1007/978-3-030-35790-0_13

which have water quality standards of 1.3 mg/L, 1.6 mg/L, and 2.0 mg/L, respectively [1, 2]. Removing these contaminants in an efficient and cost-effective manner is important for local and industrial-scale use.

Nanoparticles are of increasing interest due to their unique properties and compatibility with other materials. Applications of nanoparticles cross a variety of fields including biomedicine, with targeted drug delivery, and thermodynamics, with use in thermoelectric oxygen sensors [3–5]. Additionally, nanoparticles have been used in environmental engineering applications. The magnetic properties of magnetite nanoparticles can be exploited for use in metal contamination removal from wastewaters. Silica-coated magnetite nanoparticles have been synthesized and employed to remove heavy metals from solution [6–8]. The current research effort is engaged in developing and modifying reactors and synthesis methods in order to optimize the process of removal and recovery of metal ions in wastewater.

Experimental Procedure

Nanoparticle Synthesis

Magnetite nanoparticles with diameters ranging from 20 to 30 nm were purchased from SkySpring Nanomaterials, Inc. The nanocomposites were synthesized using a procedure developed by Dr. Edward Rosenberg at the University of Montana [7]. The synthesis involves silica coating and two surface functionalization steps. The image at left in Fig. 1 depicts the synthesis apparatus. There are four components to the setup: the mixing arm, three-neck round-bottom flask, sonicator, and nitrogen

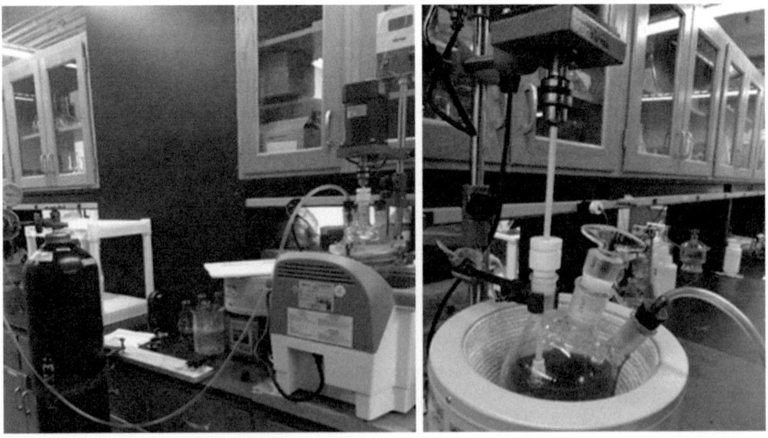

Fig. 1 Overall setup for the first step of nanoparticle synthesis (left); close up of the mixing apparatus (right)

tank, where a nitrogen atmosphere is required for the silica coating step. The image at right in Fig. 1 shows a closer view of a three-neck, round-bottom flask used for mixing and sonication of the nanoparticles throughout the process.

The first step uses tetraethyl orthosilicate (TEOS) as the source of silica for the coating and is done under a nitrogen atmosphere to prevent oxidation of the magnetite to maghemite, which could be caused by the addition of ammonium hydroxide to the nanoparticle suspension. The first surface functionalization step utilizes methyltrimethoxysilane (MTMS) and 3-chloropropyltrimethoxysilane (CPTMS), and the second functionalization step involves a poly(allylamine) (PAA) solution. PAA functionalization of the silica surface adds adsorption sites to the surface, with future work aiming to target specific metal ions using other resins and functional groups. Each step involves several hours of mixing, washes with toluene or 2-proponal or deionized (DI) water and acetone, and complete drying of the nanoparticles before beginning of the next step. The drying of the nanoparticles occurs using an air stream for evaporating as much acetone as possible before placing the nanoparticles into a vacuum chamber. Depending on the size of the container and how often the vacuum is pumped, 1 week or longer may be required to fully dry the nanoparticles. Further details can be found in *Continuous Flow Process for Recovery of Metal Contaminants from Industrial Wastewaters with Magnetic Nanocomposites* [7].

Adsorption

Adsorption experiments were performed on magnetite and synthesized magnetite nanocomposites to compare the effectiveness of synthesized nanoparticles versus natural magnetite. Each adsorption experiment was completed over a 24-hour period to determine total loading capacities. The loading capacity, q, was calculated using Eq. (1) as follows:

$$q = \frac{(C_i - C_f)V}{m} \tag{1}$$

where C_i and C_f are the initial and final concentrations, V is the total volume of solution, and m is the mass of magnetite.

The nanoparticles were suspended in a mixture of methanol (to wet the nanoparticles) and DI water. The mixture was then sonicated under vacuum to improve dispersion and remove the methanol. A separate solution was made by dissolving copper sulfate pentahydrate in DI water. After sonication, the copper solution was added and the slurry was promptly placed on a shaker table to be agitated for 24 h. Each experiment used a total volume of 250 mL and was carried out in 1000 mL flasks to allow for sufficient contact between the copper ions and the nanoparticles. After 24 h, the slurry was filtered and final copper concentration measurements were made with a Hach DR3900 spectrophotometer.

Reactor Systems

Two reactor designs, the third and fourth generation reactors, have been the focus of the research over the past year; these are shown in Figs. 2 and 3. The third generation reactor was built at pilot scale with two vertical columns. The system is operated by manual control, but an automation system capable of valve and magnet control, as well as pH and magnet temperature-data logging, has been developed. The automation system will be adapted to the fourth generation design as well. The fourth generation design features a more compact in-line electromagnet for particle collection.

The third generation design, shown at right in Fig. 2, is a dual column PVC reactor. The process outlined in the flow sheet, shown at left in Fig. 2, begins with the mixing of the nanoparticle slurry and metal bearing solution at the top and down the vertical column. The column contains static mixers which aid in maximizing contact of the metal ions with the synthesized nanoparticles. At the first branch, the slurry goes to two wye collection modules. Each module has an in-line electromagnet with core wires within the annulus to collect the nanoparticles. While in collection mode, treated effluent flows toward the center. When the magnet is de-energized, copper-loaded nanoparticles are collected and routed for further processing. The double column enables continuous processing, when one magnet is off for nanoparticle removal, the solution is diverted to the other leg.

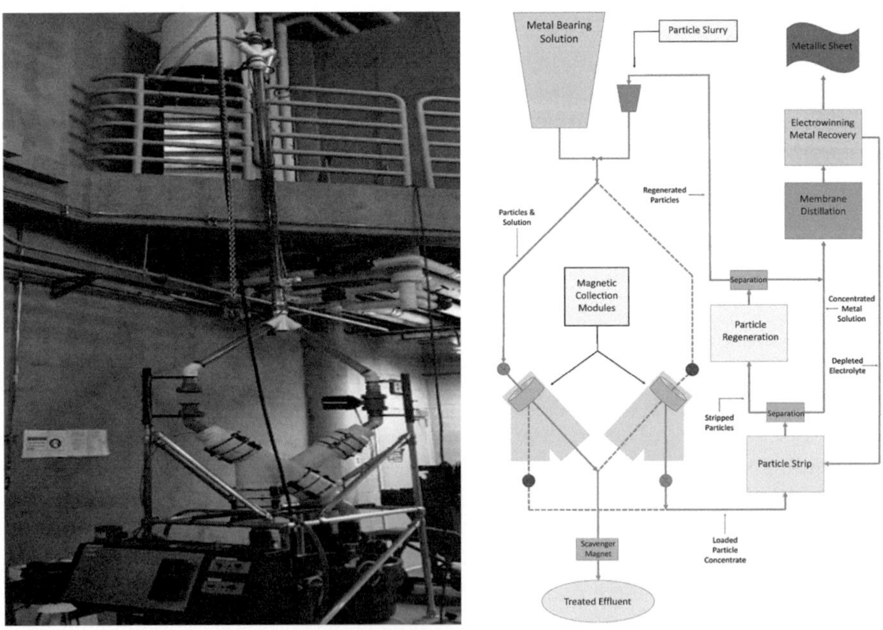

Fig. 2 Pilot-sale dual column continuous flow reactor (left). Flow diagram of the third generation dual column reactor (right) [7]

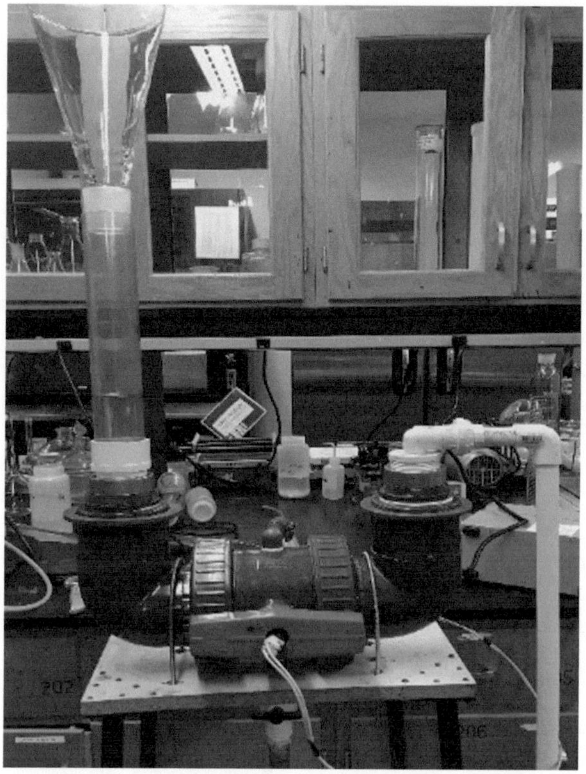

Fig. 3 The fourth generation in-line electromagnet reactor system

The fourth generation reactor, Fig. 3, uses an in-line electromagnet, placed in the horizontal section of the reactor to remove the nanoparticles from the flow. For testing, all valves were closed and the reactor was filled until water touched the bottom of the glass funnel. The power supply was turned on and the current set. The nanoparticle slurry was added to the glass funnel, with subsequent addition of water so as to keep the water level consistent. Simultaneously, the valve on the white pipe (seen in the lower right hand of Fig. 3) was opened. After a predetermined duration, in excess of the hydraulic residence time, the valve on the white pipe was closed and the power supply was shut off. The valve at the bottom of the reactor was opened for the collected particles to be gathered and weighed. Between trials, the reactor was flushed several times until clear water was observed flowing through both valves.

Results and Discussion

Nanoparticle Synthesis

Loading capacity tests were performed to determine if synthesis was successful. Figure 4 shows scanning electron microscope (SEM) images of four samples; 4a are magnetite nanoparticles purchased from SkySpring Nanomaterials, Inc. (RM–SS), 4b are magnetite nanoparticles purchased from US Research Nanomaterials, Inc. (RM–USR), 4c are silica-coated nanoparticles produced in Dr. Rosenberg's laboratory at the University of Montana (SM–ER), and 4d are silica-coated nanoparticles believed to have been oxidized during the first step of synthesis done at Montana Tech (SM–TB2). All samples were carbon coated prior to SEM imaging and energy dispersive spectroscopy (EDS) data collection.

Agglomeration of the nanoparticles is an important concern, as full dispersion results in increased surface area and increased loading on each nanoparticle. The agglomeration concern applies to commercially available nanoparticles and the extent of agglomeration after the synthesis process. As seen in Fig. 4a, b, the nanoparticles have a large size distribution ranging from tens of nanometers to over one hundred nanometers, indicating that nanoparticles directly from the suppliers are not

Fig. 4 SEM images of **a** RM-SS, **b** RM-USR, **c** SM-TB2, and **d** SM-ER

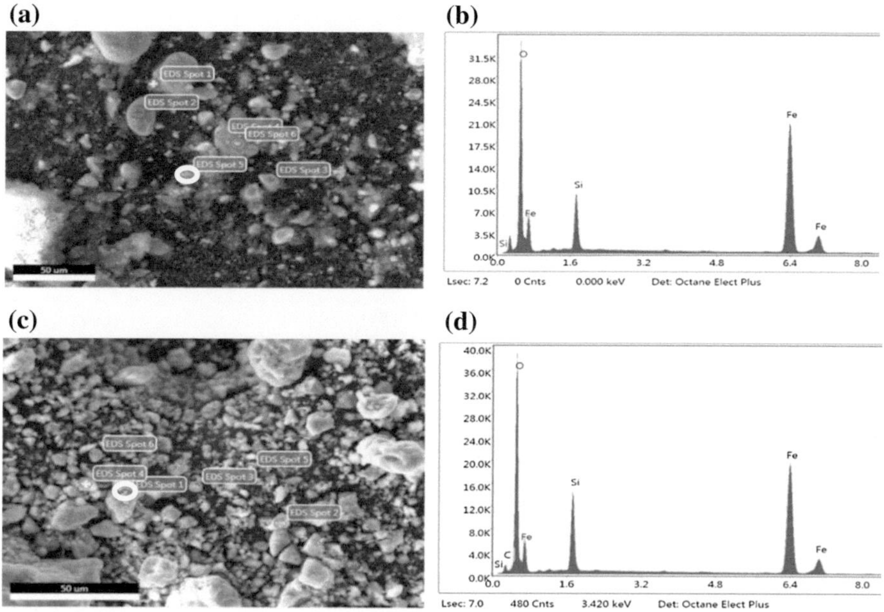

Fig. 5 **a** SEM image and **b** EDS graph of EDS spot 5 of SM-TB2 and **c** SEM image and **d** EDS graph of EDS spot 1 of SM-ER

homogeneously 20–30 nm. Figure 4c, d appear to have a lower limit on the larger end of the size spectrum, but a wide range of sizes are still present.

SEM images paired with corresponding EDS data are shown in Fig. 5. The top row represents the sample SM-TB2. The EDS Spot 5 is shown in Fig. 5b and highlighted with a circle in Fig. 5a. Similarly, the EDS Spot 1, shown in Fig. 5d, is highlighted with a circle in Fig. 5c. The bottom row represents data from the sample SM-ER. The EDS data shows primarily O, Si, and Fe in the samples. Data gathered from EDS for all spots analyzed shows several trends between the two samples in terms of weight percent of elements in the sample. SM-TB2 had overall lower oxygen content and higher iron content, whereas the silica content remained similar compared to SM-ER. The lower oxygen content in SM-TB2 does not suggest increased oxidation of the magnetite occurred during the coating process.

Adsorption

The goal of the adsorption trials is to determine trends and impact of each parameter on the system individually with eventual modelling of all parameters in order to ascertain optimal conditions under which the reactor can be operated. The modeling requires additional data, specifically with parameters that have not yet been

Table 1 Loading capacity test results

Trial	Initial concentration $\left(\frac{mg\ Cu}{L}\right)$	Fe_3O_4 (g)	Final concentration $\left(\frac{mg\ Cu}{L}\right)$	$q\left(\frac{mmol\ Cu}{g\ Fe_3O_4}\right)$
Magnetite Nanoparticles				
1	*122.47*	*0.5020*	*97.8*	*0.19*
2	*61.08*	*0.5006*	*53.6*	*0.06*
			54.8	*0.05*
3	*30.64*	*0.5005*	*24.6*	*0.05*
			25.0	*0.04*
4	15.58	0.5061	7.0	0.07
5	7.74	0.5013	2.3	0.04
Synthesized Magnetite Nanoparticles				
6	7.53	0.5003	0.07	0.06
7	15.47	0.5055	0.05	0.12
8	30.3376	0.5053	1.36	0.23
9	*46.0154*	*0.5029*	*14.44*	*0.25*
			14.55	*0.25*

extensively tested such as pH and temperature. Table 1 shows initial testing of loading capacity while changing initial concentrations and keeping the concentration of nanoparticles approximately constant.

Loading capacities of magnetite nanoparticles and synthesized magnetite nanoparticles are shown in Table 1, where trials 1, 2, 3, and 9 indicate diluted solution measurements for the final concentration. Additionally, trials 6 and 7 are not completely reliable as the final concentration measurements are outside the range of the spectrophotometer, which has a range of 0.1–8 mg/L Cu. A trend of increasing loading capacity at higher initial concentrations can be seen in both the magnetite and synthesized magnetite, with the exception of trial 4. The trend indicates that varying initial concentration of copper has a significant impact on the loading capacity of the nanoparticles, whether coated or uncoated.

Reactor Performance

The data and results discussed below are for the fourth generation design.

Optimal performance for the reactor hinges on the electromagnetic coil, which separates the loaded nanoparticles from the treated water. Thus, being able to produce a magnetic field strong enough to capture the magnetic nanoparticles is imperative to the effectiveness and efficiency of the reactor. The magnetic field strength can be calculated using a simplified equation from the Biot–Savart law in Eq. (2):

$$B = \frac{\mu_0 N I}{2R} \quad (2)$$

where B is the magnetic field strength, μ_0 is the vacuum permeability, N is the number of coil turns, I is the current, and R is the average radius.

Prior experimentation had revealed that a minimum flux density of 1 kilogauss (kG) is needed to achieve the goal of 98% collection efficiency [7]. Two coils were made for the fourth generation reactor. The coil wire was solid copper with a polyamine polymer insulation. The first coil did not produce sufficient magnetic force to attain acceptable collection efficiencies and the observed efficiencies were less than 80%. The second coil was wound using thinner wire, a reduced inner diameter, and more turns for an increased outer diameter.

A Gauss meter was used to determine the flux density of the coils, which can be seen in Table 2. The first coil did not reach the 1 kG target, which led to unsatisfactory collection efficiencies. The second coil produced higher flux densities at similar currents, as seen in Table 2, with the target of at least 1 kG being achieved with a current of 9 amps.

The second coil was used for the experiments shown in Table 3. Data collected includes nanoparticle mass, initial and final current and voltage, total time, time of brown water observed in the clean stream, and nanoparticle loss to the clean stream. The results illustrate high collection efficiencies of the prototype 3 coil in the fourth generation reactor. Collection efficiencies above 99% require a minimum of

Table 2 Coil performance

Coil version	Current (A)	Potential (V)	Flux density (kG)
Prototype 2	7	8.03	0.40
	9	10.37	0.50
	11	12.85	0.63
Prototype 3	7	29.2	0.87
	9	38.4	1.14
	9.66	42.5	1.20

Table 3 Fourth generation reactor collection efficiencies

Fe_3O_4 (g)	I_i (A)	V_i (V)	I_f (A)	V_f (V)	t_{brown} (min)	t_{total} (min)	Fe_3O_4 loss (g)	Flow rate $\left(\frac{L}{min}\right)$	Efficiency (%)
15.0015	10.00	40.45	9.45	43.41	–	1.72	0.720	2.00	95.26
15.0035	10.00	39.94	9.25	44.30	1.22	3.00	0.426	1.36	97.16
30.0406	10.00	39.79	9.15	44.82	1.50	3.00	0.265	1.36	99.12
30.0421	10.00	35.99	9.04	45.30	1.37	4.05	0.220	1.36	99.27
30.0094	9.00	36.14	9.00	44.59	1.00	4.00	0.791	1.36	97.37

10 amps and residence times average 1.4 min. The maximum field strength achieved was limited by the power supply and coil resistance.

Conclusions

Magnetite nanoparticles were synthesized for targeted loading of copper ions in solution through a three-step silica coating and functionalization process. The process is being optimized to maximize yield and minimize time required. Synthesized magnetite was shown to generally have higher loading capacities at low concentrations of copper, with maximum loading at 0.25 mmol Cu/g Fe_3O_4 for synthesized magnetite and 0.19 mmol Cu/g Fe_3O_4 for natural magnetite. A fourth generation continuous flow reactor was also evaluated. With the electromagnetic coil generating an effective magnetic field, the reactor was able to achieve over 99% efficiency in collecting nanoparticles from the water stream. Future experimentation will determine optimal operating conditions and explore additional parameters, such as pH and temperature, for impact on loading and collection.

Acknowledgements This material is based on work supported in part by the National Science Foundation EPSCoR Cooperative Agreement **OIA-1757351**. Any opinions, findings, and conclusions or recommendations expressed in this material are those of the author(s) and do not necessarily reflect the views of the National Science Foundation.

References

1. Office of Water, U.S. Environmental Protection Agency (2018) 2018 Edition of the Drinking Water Standards and Health Advisories Tables
2. Montana DEQ, Planning Prevention and Assistance Division, Water Quality Planning Bureau, Water Quality Standards Section (2012) DEQ-7 Montana Numeric Water Quality Standards. Montana Dept. of Environmental Quality, Helena, MT
3. Rettig, Frank and Ralf Moos (2010) α-Iron oxide: An intrinsically semiconducting oxide material for direct thermoelectric oxygen sensors. Sens Actuat B: Chem 145(2):685–690
4. Tadyszak Krzysztof et al (2017) Spectroscopic and magnetic studies of highly dispersible superparamagnetic silica coated magnetite nanoparticles. J Magn Magn Mater 433:254–261
5. Panyama Jayanth, Labhasetwar Vinod (2003) Biodegradable nanoparticles for drug and gene delivery to cells and tissue. Adv Drug Deliv Rev 55(3):329–347
6. Girginova Penka I et al (2010) Silica coated magnetite particles for magnetic removal of Hg2 + from water. J Colloid Interface Sci 345(2):234–240
7. Hutchins D (2018) Continuous flow process for recovery of metal contaminants from industrial wastewaters with magnetic nanocomposites. PhD thesis, Montana Technological University
8. Hughes M et al (2006) Silica polyamine composites: New supramolecular materials for cation and anion recovery and remediation. Macromol Symp 235:161–178

Author Index

T. S. Srivatsan and M. Gupta (eds.), *Nanocomposites VI: Nanoscience and Nanotechnology in Advanced Composites*, The Minerals, Metals & Materials Series, https://doi.org/10.1007/978-3-030-35790-0

Subject Index

T. S. Srivatsan and M. Gupta (eds.), *Nanocomposites VI: Nanoscience and Nanotechnology in Advanced Composites*, The Minerals, Metals & Materials Series, https://doi.org/10.1007/978-3-030-35790-0

MIX
Papier aus verantwortungsvollen Quellen
Paper from responsible sources
FSC® C105338

If you have any concerns about our products,
you can contact us on
ProductSafety@springernature.com

In case Publisher is established outside the EU,
the EU authorized representative is:
Springer Nature Customer Service Center GmbH
Europaplatz 3, 69115 Heidelberg, Germany

Printed by Libri Plureos GmbH
in Hamburg, Germany